AS

VISUAL
REVISION
GUIDE

SUCCESS

CHEMISTRY

Bob McDuell

Contents

Basic concepts

Physical chemistry

Inorganic chemistry

Organic chemistry

States of matter

- There are three states of matter: solids, liquids and gases.
- Energy is required to turn solids into liquids and liquids into gases.

Solids, liquids and gases

The diagrams show arrangements of particles in solids, liquids and gases.

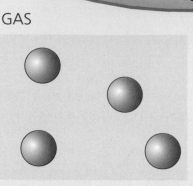
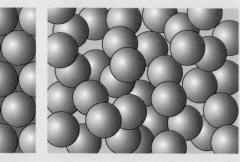

SOLID   LIQUID   GAS

State of matter	Arrangement of particles	Motion of particles	Forces between particles
solid	particles are regularly arranged in crystalline structure	vibration	usually very strong forces between particles
liquid	particles are irregularly spaced	slow, random movement	strong forces between particles
gas	particles are irregularly and widely spaced	rapid, random motion	no forces between particles in an **ideal gas** but very weak forces in real gases

Changes of state

Energy is needed to turn a solid into a liquid and, again to turn a liquid into a gas.

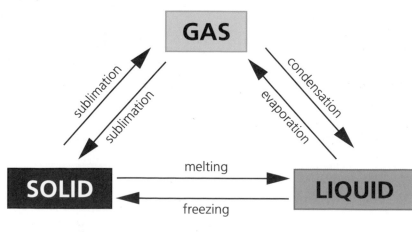

Changes in the states of matter

Changes of state continued

Enthalpy of fusion
This is the energy required to change 1 mole of a solid into 1 mole of a liquid at the melting point.

Enthalpy of vaporisation
This is the energy required to change 1 mole of liquid to 1 mole of gas at the boiling point.

Other ways of breaking up a solid lattice
Melting can break down a solid structure, such as sodium chloride. The energy supplied breaks up the structure.

A solid structure can also be broken down by dissolving in water. The charges within the water molecules pulls apart the sodium chloride lattice. Water molecules then surround the ions in solution.

Relationship between boiling points of liquids

For most liquids there is a simple relationship between boiling point and the molar energy change for evaporation. For most liquids, when the molar energy change for evaporation is plotted against boiling point (in Kelvin), the points fall on a straight line.

This is expressed in Trouton's Rule, which states that the molar enthalpy change of vaporisation of a liquid, at its normal boiling point, divided by its boiling point in Kelvin, is constant.

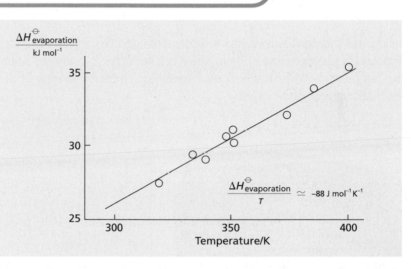

$$\frac{\Delta H^{\ominus}_{evaporation}}{T} \simeq -88 \text{ J mol}^{-1}\text{K}^{-1}$$

Quick test

1 The graph shows the energy required to turn one mole of different liquids to a gas at the boiling point.

a What name is given to the energy required for this change?

b For most liquids there is a pattern between this energy and the boiling point. What is this pattern?

c There are three important exceptions. Which substances are these?

2 Explain why energy is needed to change a liquid to a gas and a solid to a liquid.

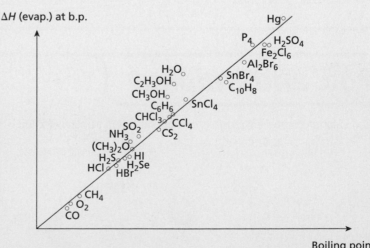

1. (a) The enthalpy of vaporisation; (b) The energy required is directly proportional to the boiling point (on the Kelvin scale); (c) Ethanol, methanol and water.   2. At each change, particles have to be moved apart. There are attractive forces holding them together. Also, the particles have to be given additional energy to enable them to move faster.

5

Compound formation

- Compounds are formed when two or more elements combine.
- Forming a compound from its elements is called <u>synthesis</u>.

Synthesis

The table compares properties of a mixture of elements and the compound formed when the elements combine.

Mixture of elements	Compound
the percentage of each element present is variable (not fixed)	contains a fixed percentage of each element
the properties of the mixture are those of the constituent elements	the properties of a compound are different from its constituent elements

A mixture of hydrogen and oxygen is a colourless gas. The compound of hydrogen and oxygen, water, is a liquid at room temperature. Water has entirely different properties from both hydrogen and oxygen.

The diagram shows what happens when a compound is formed when hydrogen and oxygen combine and form water.

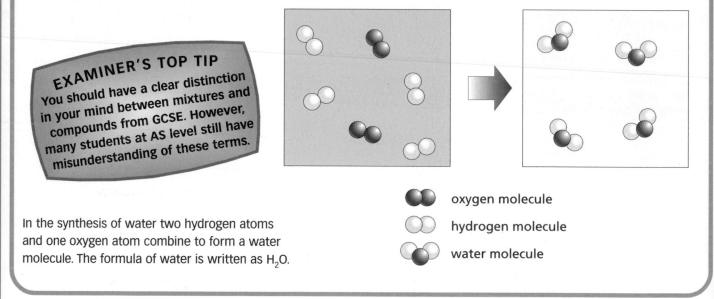

oxygen molecule

hydrogen molecule

water molecule

EXAMINER'S TOP TIP
You should have a clear distinction in your mind between mixtures and compounds from GCSE. However, many students at AS level still have misunderstanding of these terms.

In the synthesis of water two hydrogen atoms and one oxygen atom combine to form a water molecule. The formula of water is written as H_2O.

Iron and sulphur

Iron and sulphur combine to form a compound called iron(II) sulphide.

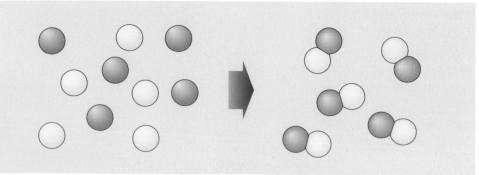

The synthesis of iron(II) sulphide

Naming compounds

- Many compounds have trivial names that do not clearly show which elements they contain.
  e.g. water H_2O
       methane CH_4
       ammonia NH_3
       benzene C_6H_6

- Sometimes the name indicates the elements present and the number of each type of atom.
  e.g. carbon monoxide CO
       carbon dioxide CO_2
       nitrogen trichloride NCl_3
       carbon tetrachloride CCl_4

- Where a compound contains a metal, the metal is written first.
  e.g. calcium oxide CaO
       calcium chloride $CaCl_2$

- A compound formed when two elements combine is given a name ending in -ide.
  e.g. calcium oxide CaO
       sodium chloride $NaCl$
  Exception:
  sodium hydroxide $NaOH$
  This has a name ending in -ide but contains three elements: sodium, hydrogen and oxygen.

- Compounds with names ending in -ate and -ite contain oxygen.
  e.g. sodium nitrate $NaNO_3$
       sodium nitrite $NaNO_2$
       sodium carbonate Na_2CO_3
       sodium sulphate Na_2SO_4
       sodium sulphite Na_2SO_3

N.B. A compound with an ending -ate contains more oxygen than a compound ending in -ite.

- A prefix per- indicates that the compound contains extra oxygen.
  e.g. water (hydrogen oxide) H_2O
       hydrogen peroxide H_2O_2

- A compound with a prefix thio- contains one less oxygen atom and one more sulphur atom.
  e.g. sodium sulphate Na_2SO_4
       sodium thiosulphate $Na_2S_2O_3$

- A compound may contain a roman numeral within its name. This shows the oxidation state of the element.
  e.g. copper(II) oxide CuO
  Copper is in oxidation state +2.

EXAMINER'S TOP TIP
Understanding the rules for naming compounds will help you when naming compounds, understanding reactions and writing equations.

Quick test

1 Finish the diagram to show what happens when nitrogen and hydrogen combine to form ammonia, NH_3.

       nitrogen molecule
       hydrogen molecule

2 When potassium nitrate is heated, potassium nitrite is formed. What other product is formed?

3 Nitrogen forms a number of different oxides, including

    nitrogen monoxide   dinitrogen tetroxide
    nitrogen dioxide   dinitrogen pentoxide

Give the formula for each of these compounds.

4 Which of these compounds contain oxygen?

potassium chlorate   potassium hydroxide
potassium oxide   potassium hydride
potassium chloride

5 Sodium oxide has a formula of Na_2O. Write down the formula of sodium peroxide.

6 The table gives masses of sulphur and oxygen combined to give sulphur dioxide.

Which one does not contain pure sulphur dioxide?

mass of sulphur in g	mass of oxygen in g
1.0	1.0
1.5	1.5
1.7	2.0
2.0	2.0

Explain your choice.

2. oxygen 3. NO, NO_2, N_2O_4, N_2O_5 4. potassium chlorate, potassium oxide, potassium hydroxide 5. Na_2O_2 6. 1.7g of sulphur and 2.0g of oxygen. In all other cases, the mass of oxygen combined is the same as the mass of oxygen.

Types of chemical reaction

There are many different types of chemical reaction.

- Decomposition is the splitting up of a compound. This can be done in different ways.
- Oxidation and reduction are opposite processes. Oxidation involves addition of oxygen, loss of hydrogen or loss of electrons.

Thermal decomposition

Thermal decomposition is the splitting up of a substance by heating.
For example, zinc carbonate decomposes on heating into zinc oxide and carbon dioxide.

$$ZnCO_3(s) \longrightarrow ZnO(s) + CO_2(g)$$

Oxidation and reduction

An oxidation reaction can be simply defined as a reaction where oxygen is added. Reduction is the opposite of oxidation and involves oxygen being taken away. The table gives examples of oxidation and reduction. There are other definitions of oxidation and reduction. An oxidation reaction is also a reaction where hydrogen is lost or electrons are removed. Reduction can also be described as a reaction where hydrogen is gained or electrons are added.

oxidation	reduction
combustion	extracting metals from metal ores
corrosion	
respiration	

A sodium ion gains an electron and so is reduced.
A chloride ion loses an electron and so is oxidised.
The table gives common oxidising and reducing agents.

Common oxidising agents	Common reducing agents
oxygen	hydrogen
chlorine	sulphur dioxide
potassium manganate(VII)	metals
potassium dichromate(VI)	hydrogen peroxide
hydrogen peroxide	

e.g. 1: Lead(II) oxide changed to lead by heating with carbon.

$$PbO(s) + C(s) \longrightarrow Pb(s) + CO(g)$$

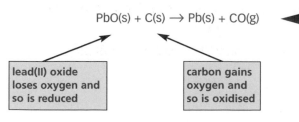

lead(II) oxide loses oxygen and so is reduced

carbon gains oxygen and so is oxidised

In this reaction, carbon brings about the reduction of the lead(II) oxide, so the carbon is the **reducing agent**. The lead(II) oxide provides the oxygen that enables the carbon to be oxidised and so is called the **oxidising agent**.

e.g. 2: Electrolysis of molten sodium chloride produces sodium and chlorine.

$$Na^+ + e^- \longrightarrow Na$$
$$Cl^- \longrightarrow \tfrac{1}{2}Cl_2 + e^-$$

Acid–base reactions

Acids and bases react together to form salts.

Acids are compounds with replaceable hydrogen atoms. Dissolved in water they produce $H^+(aq)$ ions

Bases are metal oxides or hydroxides

Alkalis are soluble bases containing OH^- ions

Acids can be classified as **strong** or **weak** acids. A strong acid is completely ionised and a weak acid is only partially ionised.

e.g. Nitric acid is completely ionised

$$HNO_3(g) + water \rightarrow H^+(aq) + NO_3^-(aq)$$ Strong acid

Ethanoic acid is only partially ionised

$$CH_3CO_2H(l) + water \rightleftharpoons CH_3CO_2^-(aq) + H^+(aq)$$ Weak acid

A **neutralisation** reaction takes place when a base or alkali reacts with an acid to form a salt and water only.

e.g. Sulphuric acid and sodium hydroxide $H_2SO_4(aq) + 2NaOH(aq) \rightarrow Na_2SO_4(aq) + 2H_2O(l)$

This reaction can be summarised by the ionic equation

$$H^+(aq) + OH^-(aq) \rightarrow H_2O(l)$$

Ionic precipitation

A **precipitate** is a solid formed when two solutions are mixed together.

e.g. Silver chloride is formed when silver nitrate and hydrochloric acid solutions are mixed.

$$AgNO_3(aq) + HCl(aq) \rightarrow AgCl(s) + HNO_3(aq)$$

When this is written in the form of ions:

$$Ag^+(aq) + NO_3^-(aq) + H^+(aq) + Cl^-(aq) \rightarrow AgCl(s) + H^+(aq) + NO_3^-(aq)$$

Removing ions that appear on both sides
(called **spectator ions**) gives:

$$Ag^+(aq) + Cl^-(aq) \rightarrow AgCl(s)$$

Quick test

1 Name a substance that can act as an oxidising agent or a reducing agent depending upon conditions.

2 Write an ionic equation for the precipitation of barium sulphate when barium nitrate and sulphuric acid are mixed.

3 The equation for the reaction of manganese(IV) oxide and concentrated hydrochloric acid is shown below.

$$MnO_2(s) + 4HCl(aq) \rightarrow MnCl_2(aq) + 2H_2O(l) + Cl_2(g)$$

a Which substance is oxidised and which is reduced?

b Which substance is the oxidising agent and which the reducing agent?

4 Write down the names of three strong acids.

5 Lead(IV) oxide splits up on heating. $2PbO_2(s) \rightarrow 2PbO(s) + O_2(g)$ Which type of reaction is this?

1. Hydrogen peroxide 2. $Ba^{2+}(aq) + SO_4^{2-}(aq) \rightarrow BaSO_4(s)$ 3. (a) HCl is oxidised and MnO_2 is reduced (b) HCl is the reducing agent and MnO_2 is the oxidising agent 4. nitric acid, sulphuric acid and hydrochloric acid 5. thermal decomposition

Working out the formula of a compound

A chemical compound has a fixed chemical composition.
It is possible to work out the formula of a compound from its chemical composition.
You will need to use relative atomic masses on the data sheet (page 90).

Working out the formula of a compound – 1

Here are two examples of how to work out the formula of a compound from **reaction masses**.

1 Magnesium oxide
The formula of magnesium oxide can be found by burning a known mass of magnesium ribbon in a crucible with a lid.

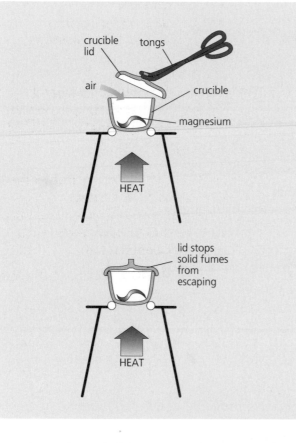

Sample results might be:

  a Mass of crucible + lid   =   25.15 g
  b Mass of crucible, lid and magnesium   =   25.27 g
     Mass of magnesium **b – a**   =   0.12 g
  c Mass of crucible, lid and
     magnesium oxide formed   =   25.35 g
     Mass of magnesium oxide **c – a**   =   0.20 g

From these results
   0.12 g of magnesium combines with (0.20 – 0.12) g
   of oxygen to form 0.20 g of magnesium oxide.
   0.12 g of magnesium combines with 0.08 g of oxygen.

Using the relationship   no. of moles = $\dfrac{\text{mass of substance}}{\text{mass of 1 mole}}$

$\dfrac{0.12}{24}$ mole of magnesium atoms combine with $\dfrac{0.08}{16}$ mole of oxygen atoms.

0.005 mole of magnesium atoms combine with 0.005 mole of oxygen atoms.

(1 mole of magnesium atoms contains the same number of atoms as 1 mole of oxygen atoms).

The simplest formula is, therefore, **MgO**.

Good results can be obtained by doing a series of experiments with different masses of magnesium ribbon and recording the results on a graph.

This procedure reduces the inaccuracy of individual experiments.

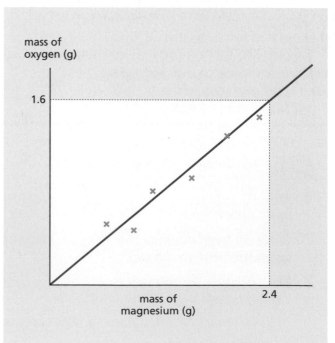

2 Copper(I) oxide

A known mass of copper(I) oxide is reduced to copper using hydrogen in apparatus shown in the diagram.

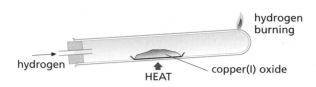

Sample results might be:

a Mass of combustion boat   =   12.20 g
b Mass of boat + copper(I) oxide   =   13.64 g
    Mass of copper(I) oxide **b – a**   =   1.44 g

c   Mass of boat + copper
    after reduction   =   13.48 g
    Mass of copper **c – a**   =   1.28 g

1.28 g of copper combines with (1.44 – 1.28) g oxygen to form 1.44 g of copper(I) oxide.

1.28 g of copper combines with 0.16 g of oxygen.

$\dfrac{1.28}{64}$ mole of copper atoms combine with $\dfrac{0.16}{16}$ mole of oxygen atoms.

0.02 mole of copper atoms combine with 0.01 mole of oxygen atoms.

Simplest formula is **Cu_2O**.

Working out the formula of a compound – 2

Here is how you work out the formula of a compound from its **percentage composition**. A compound contains 75% carbon and 25% hydrogen. Calculate the empirical (or simplest) formula of this compound.

		C	H
a	Percentage composition	75	25
b	Mass of 1 mole of atoms	12	1
	a ÷ b	$\dfrac{75}{12} = 6.25$	$\dfrac{25}{1} = 25$
	÷ smallest	1	4

Empirical or simplest formula = **CH_4**

To work out the molecular formula you need the mass of 1 mole of the compound. The mass of 1 mole of the compound is 16.

$(12 + 4 \times 1)n = 16$; $n = 1$

Molecular formula is CH_4

Empirical formula

When a formula is worked out from combining masses or from percentages, the formula is called an **empirical** or **simplest formula**.

For example, the empirical formula CH_2 shows that carbon and hydrogen combine in the ratio of 6:1. The actual formula could be CH_2 or some multiple of it.

e.g. C_2H_4 C_3H_6 C_4H_8 etc.

The molecular formula can be found if the mass of 1 mole (RFM) is known.

If the mass of 1 mole of a compound with empirical formula CH_2 is known to be 28, the molecular formula is C_2H_4.

Quick test

Remember the mass of 1 mole = RAM in g.

1 *Calculate the simplest formulae from the following data:*

   a   6 g of carbon combine with 1 g of hydrogen,

   b   0.7 g of nitrogen combines to form 1.5 g of nitrogen oxide,

   c   4.14 g of lead combines with 0.64 g of oxygen to form a lead oxide,

   d   0.02 g of hydrogen combines with 0.32 g of oxygen to form a hydrogen oxide,

   e   1.12 g of iron combines with oxygen to form 1.60 g of iron oxide.

2 *In each case, calculate the empirical (simplest) formula from the percentages given.*

   a   Sulphur 50%, oxygen 50%.

   b   Sulphur 40%, oxygen 60%.

   c   Carbon 84%, hydrogen 16%.

   d   Carbon 40%, hydrogen 6.67%, oxygen 53.33%.

   e   Iron 36.8%, sulphur 21.1%, oxygen 42.1%.

1. (a) CH_2; (b) NO; (c) PbO_2; (d) H_2O; (e) Fe_2O_3.   2. (a) SO_3; (b) SO_3; (c) C_7H_{16}; (d) CH_2O; (e) $FeSO_4$

Writing chemical equations

- A chemical equation is a shorthand way of representing a chemical reaction.
- They are recognised by chemists all round the world.
- Here is part of an A-level textbook in Malay. Notice that, even if you cannot read Malay, you can recognise the chemical reaction being referred to.

> Hitungkan jisim Plumbum(II) nitrat yang mesti dipanaskan untuk menghasilkan 2.3 g plumbum(II) oksida. Persamaan tindak balas:
>
> $$2Pb(NO_3)_2 \rightarrow 2PbO + 4NO_2 + O_2$$

- It is always theoretically possible to obtain the equation from the results of an experiment.

EXAMINER'S TOP TIP
You should take every opportunity in examination papers to write balanced symbol equations.

Steps in writing a chemical equation

1. Write down the equation as a word equation using either the information given or your memory. Include all reacting substances (reactants) and all products (do not forget small molecules such as water):
   e.g. calcium hydroxide + hydrochloric acid → calcium chloride + water.

2. Fill in the correct formulae for all the reacting substances and products:
   $Ca(OH)_2 + HCl \rightarrow CaCl_2 + H_2O$

3. Now balance the equation. During a chemical reaction, atoms cannot be created or destroyed (Law of conservation of mass). There must be the same total numbers of the different atoms before and after the reaction. When balancing an equation only the **proportions** of the reacting substances and products can be altered – not the formulae.
   $Ca(OH)_2 + 2HCl \rightarrow CaCl_2 + 2H_2O$

Using state symbols

The states of reacting substances and products can be included in small brackets after the formulae in equations:

- **(s) for solid (sometimes (c) is seen for crystalline solid)**
- **(l) for liquid**
- **(g) for gas**
- **(aq) for an aqueous solution where water is the solvent**

These state symbols may not be required and may not give you extra credit from the examiner but they do help your thinking considerably and give a good impression.

$$Ca(OH)_2(aq) + 2HCl(aq) \rightarrow CaCl_2(aq) + 2H_2O(l)$$

Ionic equations

Ionic equations are useful because they emphasise the important changes taking place in a chemical reaction.

For the reaction between acidified potassium manganate(VII) and iron(II) sulphate the full equation is:

$$2KMnO_4 + 8H_2SO_4 + 10FeSO_4 \rightarrow 2MnSO_4 + 8H_2O + 5Fe_2(SO_4)_3 + K_2SO_4$$

In this equation there are a number of ions (spectator ions) which take no part in the reaction. They are present before and after the reaction. These tend to confuse the change taking place and can often be ignored. Expanding the equation to show the ions present gives:

$$2(K^+MnO_4^-) + 8(2H^+SO_4^{2-}) + 10(Fe^{2+}SO_4^{2-}) \rightarrow 2(Mn^{2+}SO_4^{2-}) + 8H_2O + 5(2Fe^{3+}3SO_4^{2-}) + 2K^+ + SO_4^{2-}$$

Crossing out ions appearing on both sides of the equation gives:

$$2MnO_4^- + 16H^+ + 10Fe^{2+} \rightarrow 2Mn^{2+} + 8H_2O + 10Fe^{3+}$$

Half equations

The equation can be divided through by 2 to give the simplest equation.

$$MnO_4^- + 8H^+ + 5Fe^{2+} \rightarrow Mn^{2+} + 4H_2O + 5Fe^{3+}$$

This represents the simplest ionic equation for this reaction. The equation must be balanced in the usual way with equal numbers of each type of atom on the left- and right-hand sides. In addition, the algebraic sum of the charges on the left-hand side equals that of the right-hand side.

For the example:
LHS $(-1) + 8(+) + 5(2+) = 17(+)$
RHS $(2+) + 4(0) + 5(3+) = 17(+)$

This ionic equation can be broken down into two half equations, which added together produce the overall equation:

1. $MnO_4^- + 8H^+ + 5e^- \rightarrow Mn^{2+} + 4H_2O$

2. $5Fe^{2+} \rightarrow 5Fe^{3+} + 5e^-$

Equation 1. represents the reduction of manganese(VII) to manganese(II). Equation 2. represents the oxidation of iron(II) to iron(III).

A list of common half equations and their use in building up ionic equations is given on page 91.

Quick test

1 *Balance the following equations*

  a $CaCO_3 + HNO_3 \rightarrow Ca(NO_3)_2 + H_2O + CO_2$

  b $CuO + HCl \rightarrow CuCl_2 + H_2O$

  c $NH_3 + O_2 \rightarrow N_2 + H_2O$

  d $Cu + HNO_3 \rightarrow Cu(NO_3)_2 + NO + H_2O$

  e $CH_4 + O_2 \rightarrow CO_2 + H_2O$

2 *Write the simplest ionic equations for 1a and 1b. Include state symbols.*

Using equations for calculations

- Chemical equations can be used to summarise chemical reactions.
- They can also be used to calculate quantities of materials that react together and quantities of materials that can be produced.
- It is important to manufacturers and users of chemicals to know how much is needed and how much is likely to be produced in a reaction.

Calculating the mass of product formed

The reaction that occurs when calcium carbonate is heated can be summarised by a chemical equation.

$$CaCO_3(s) \rightarrow CaO(s) + CO_2(g)$$

EXAMINER'S TOP TIP

Avoid making simple arithmetical mistakes here. Before moving on, always check the sum of the reactants and the products. They must be the same.

Using relative atomic masses it is possible to work out the relative masses of the different chemicals involved.

(Relative atomic masses in g: Ca = 40, O = 16, C = 12)

$$\{40 + 12 + (3 \times 16)\} \rightarrow \{40 + 16\} + \{12 + (2 \times 16)\}$$

| 100 | 56 | 44 |

Here both sides of the equation add up to 100.

Now the equation tells us that 100 g of calcium carbonate would produce 56 g of calcium oxide.

Industrial scale

A manufacturer producing calcium oxide on a very large scale from limestone would know that 1 tonne of limestone would produce 0.56 tonnes of calcium oxide.

Calcium carbonate and hydrochloric acid

Now consider the reaction of calcium carbonate with hydrochloric acid.
The equation for the reaction is:

$$CaCO_3(s) + 2HCl(aq) \rightarrow CaCl_2(aq) + H_2O(l) + CO_2(g)$$

Using the relative atomic masses gives:

100 g of calcium carbonate reacts with 73 g of hydrochloric acid to produce 111 g of calcium chloride, 18 g of water and 44 g of carbon dioxide.

2.00 g of calcium carbonate would produce 2.22 g of calcium chloride.

Calculating the quantities of reactants

What is the quantity of sulphur needed to react with 1.0 g of iron to make iron(II) sulphide?

First you need the equation.

$$Fe(s) + S(s) \rightarrow FeS(s)$$

Then using the relative atomic masses in g: Fe = 56, S = 32

56 g of iron react with 32 g of sulphur to produce 88 g of iron(II) sulphide.

1.0 g of iron would require $\frac{32}{56} \times 1.0$ g of sulphur

= 0.57 g of sulphur.

Calculating the yield from a blast furnace

On a larger scale a steel manufacturer may have 100 tonnes of iron from the blast furnace and it contains 0.1% silicon(IV) oxide. He wants to add limestone to a furnace to remove this.

It is important that the right amount is added.

If not enough is added, silicon(IV) oxide will remain in the furnace. Adding too much is wasteful and increases the impurities in the furnace.

Calculating the mass of slag

How much silicon(IV) oxide is in the furnace?

0.1% of 100 tonnes = 0.1 tonnes.

The equation for the reaction of silicon(IV) oxide and limestone is:

$$CaCO_3(s) + SiO_2(l) \rightarrow CaSiO_3(l) + CO_2(g)$$

Using relative atomic masses in g
Ca = 40, C = 12, O = 16, Si = 28
100 tonnes of calcium carbonate reacts with 60 tonnes of silicon(IV) oxide to form 116 tonnes of slag.
0.1 tonnes of calcium carbonate reacts with 0.06 tonnes of silicon(IV) oxide to form 0.116 tonnes of slag.

In these calculations the quantities are theoretical. In practice, because of other factors, e.g. side reactions, less may be produced or more may be needed than these calculations suggest.

Percentage yields

The yield obtained in a chemical reaction is usually expressed as a percentage.

$$\text{percentage yield} = \frac{\text{actual yield}}{\text{theoretical yield}} \times 100$$

Barium sulphate can be prepared by reacting barium nitrate and sodium sulphate.

$$Ba(NO_3)_2(aq) + Na_2SO_4(aq) \rightarrow BaSO_4(s) + 2NaNO_3(aq)$$

Suppose a sample of 2.61g of barium nitrate is converted into barium sulphate and 1.65 g of the dry solid is produced.
Actual yield = 1.65 g
Calculate the theoretical yield.
Mass of 1 mole of barium nitrate = {137 + (2 × 14) + (6 × 16)} = 261 g

Number of moles of barium nitrate used = $\frac{2.61}{261}$ = 0.01

Number of moles of barium sulphate if yield 100% = 0.01
Mass of 1 mole of barium sulphate = {137 + 32 + (4 × 16)} = 233 g
Theoretical yield of barium sulphate = 233 × 0.01 = 2.33 g
Calculate the percentage yield.

$$\begin{aligned}\text{percentage yield} &= \frac{\text{actual yield}}{\text{theoretical yield}} \times 100 \\ &= \frac{1.65}{2.33} \times 100 \\ &= 70.8\%\end{aligned}$$

EXAMINER'S TOP TIP
The answer is given to 3 significant figures as the data given was to 3 significant figures.

Quick test

Relative atomic masses will be found on page 91.

1 Calculate the mass of potassium carbonate formed when 2.00 g of potassium hydrogencarbonate is heated until it has a constant mass.

$$2KHCO_3(s) \rightarrow K_2CO_3(s) + H_2O(l) + CO_2(g)$$

2 What mass of oxygen is needed to combine with 2.0 g of magnesium?

$$2Mg(s) + O_2(g) \rightarrow 2MgO(s)$$

3 In a chemical reaction the actual yield is 8.00 g and the theoretical yield is 12.0 g. What is the percentage yield?

1. 1.38 g 2. 1.33 g 3. 00.6%

Reacting volumes of gases

- Gases have very low densities and as a result a small volume of gas will have a small mass.
- It is much more convenient to measure volumes of gases rather than masses.
- However, the volume of a gas changes as temperature and pressure change.

Avogadro's Hypothesis

This states that:

equal volumes of gases under the same conditions of temperature and pressure contain the same number of molecules.

At room temperature (25 °C or 298K) and 100 kPa pressure (atmospheric pressure), 1 mole of any gas occupies 24 dm^3 or 24 000 cm^3.

Calculating the number of moles of gas

Assuming room temperature and atmospheric pressure (r.t.p.), the number of moles of gas, n, can be calculated using the expressions

$$n = \frac{\text{volume (in dm}^3\text{)}}{24} \quad \text{or} \quad n = \frac{\text{volume (in cm}^3\text{)}}{24\,000}$$

e.g. Calculate the number of moles of hydrogen in 12 dm^3 at r.t.p.

$$n = \frac{12}{24} = 0.5 \text{ mol}$$

The Ideal Gas Equation

The Ideal Gas Equation can be used for calculations involving gases.

$$pV = nRT$$

where p is pressure in kPa
V is volume in m^3
n is number of moles of gas or mass/molar mass
T is temperature in K
R is the gas constant = 8.31 JK^{-1}mol^{-1}

e.g. Calculate the mass of nitrogen which exerts a pressure of 2.5 × 10^5 Pa on a 300 cm^3 vessel at 20 °C (N = 14, R = 8.31 JK^{-1}mol^{-1}).

$pV = nRT$ and $n = \frac{m}{M}$ where m = mass and M = mass of 1 mol

$$pV = \frac{mRT}{M}$$

$$m = \frac{2.5 \times 10^5 \times 300 \times 28}{10^6 \times 8.31 \times 293}$$

$$= 0.862 \text{ g}$$

EXAMINER'S TOP TIP

Students always have problems with units here.
To convert cm^3 into m^3 multiply by 10^{-6}.
To convert dm^3 into m^3 multiply by 10^{-3}.
To convert kPa into Pa multiply by 10^3.
To convert °C into K add 273.

Reacting volumes of gases

Hydrogen and chlorine combine to form hydrogen chloride.

$$H_2(g) + Cl_2(g) \rightarrow 2HCl(g)$$

When one volume of hydrogen and one volume of chlorine combine, two volumes of hydrogen chloride are formed.
This is providing the volumes of gases are measured under the same conditions of temperature and pressure.
When gases react they do so in simple ratios by volume, providing temperature and pressure are unchanged.

EXAMINER'S TOP TIP
You must be careful here and don't fall into the trap of thinking that the sum of the volumes of the reactants is equal to the sum of the volumes of the products. It is true in this case but it is unusual.

Burning carbon monoxide in air

$$2CO(g) + O_2(g) \rightarrow 2CO_2(g)$$

The equation shows that two volumes of carbon monoxide combines with one volume of oxygen to produce two volumes of carbon dioxide.
So under the same conditions of temperature and pressure,
100 cm³ of carbon monoxide reacts with 50 cm³ of oxygen to produce 100 cm³ of carbon dioxide

Reactions where water is a product

Remember that the volumes of gases are often measured at room temperature and atmospheric pressure.
Under these conditions any steam produced will condense and form liquid water.
The volume of this water is negligible compared to the volumes of gases.
(On condensation water has less than one thousandth of the volume of the same mass of steam.)
In the combustion of hydrogen:

$$2H_2(g) + O_2(g) \rightarrow 2H_2O(l)$$

Two volumes of hydrogen combine with one volume of oxygen but the volume of the products is zero as the water will have condensed.

Quick test

1 Calculate the volume of 1.1 g of carbon dioxide, CO_2, at room temperature and atmospheric pressure (r.t.p.).

2 What is the mass of 4.8 dm³ of sulphur dioxide, SO_2, at r.t.p.?

3 The equation for the combustion of methane in excess air is

$$CH_4(g) + 2O_2(g) \rightarrow CO_2(g) + 2H_2O(l)$$

 a What volume of oxygen is needed to burn 50 cm³ of methane?

 b What volume of products is formed?

 (All volumes measured at r.t.p.)

Titrations

- A procedure where two solutions are mixed until exactly reacting quantities of the two solutions are present is called a <u>titration</u>.
- The point where exactly reacting quantities are present is called the <u>equivalence</u> (or <u>end</u>) <u>point</u>.
- End points are often found using <u>indicators</u>.
- Some titrations involve acids and bases. These are called <u>acid–base titrations</u>.

Measuring solution concentrations

The concentration of solutions can be measured in units of **g dm^{-3}** or **g/dm^3**.

For example: 1.0 g of sodium hydroxide is added to water and made up to 250 cm^3.

What is the concentration of the solution?

$$\text{Concentration} = \frac{1.0 \times 1000 \text{ g dm}^{-3}}{250}$$

$$= 4.0 \text{ g dm}^{-3}$$

Another way of measuring the concentration of a solution is called **molarity**.
This involves units of **mol dm^{-3}**.
<u>Solutions of the same molarity contain the same number of particles.</u>

Calculate the molarity of a solution containing 0.84 g of sodium hydrogencarbonate made up to 100 cm^3 of solution.

Mass of 1 mole of NaHCO$_3$ = 23 + 1 + 12 + (3 × 16)   = 84 g

Number of moles of NaHCO$_3$ in 0.84 g sample = $\frac{0.84}{84}$ = 0.01

0.01 mol of NaHCO$_3$ is dissolved to make 100 cm^3 of solution.
Molarity of solution = 0.1 mol dm^{-3}

Stages in making a solution of known concentration

A solution of known concentration is also called a **standard solution**.

| Weigh out the required mass of solid | Add the solid to water. Stir to dissolve solid | Make up to solution to 250cm^3 with distilled water |

Acid–base titrations

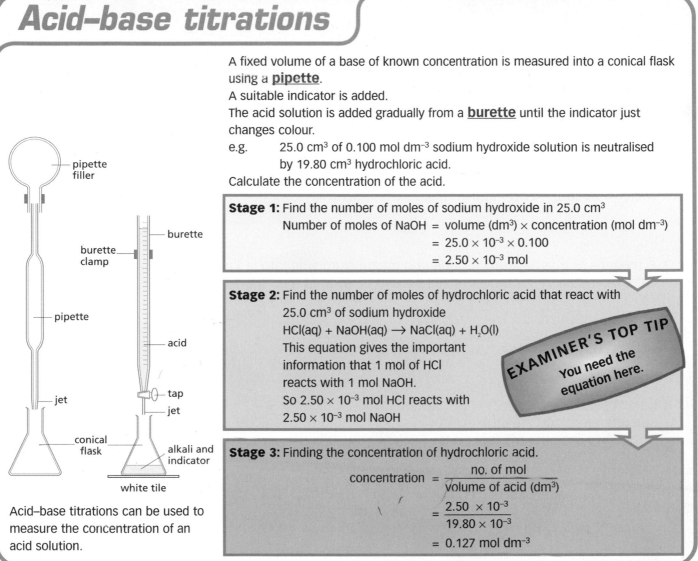

A fixed volume of a base of known concentration is measured into a conical flask using a **pipette**.

A suitable indicator is added.

The acid solution is added gradually from a **burette** until the indicator just changes colour.

e.g.   25.0 cm³ of 0.100 mol dm⁻³ sodium hydroxide solution is neutralised by 19.80 cm³ hydrochloric acid.

Calculate the concentration of the acid.

Stage 1: Find the number of moles of sodium hydroxide in 25.0 cm³

Number of moles of NaOH = volume (dm³) × concentration (mol dm⁻³)

= 25.0 × 10⁻³ × 0.100

= 2.50 × 10⁻³ mol

Stage 2: Find the number of moles of hydrochloric acid that react with 25.0 cm³ of sodium hydroxide

$HCl(aq) + NaOH(aq) \rightarrow NaCl(aq) + H_2O(l)$

This equation gives the important information that 1 mol of HCl reacts with 1 mol NaOH.

So 2.50 × 10⁻³ mol HCl reacts with 2.50 × 10⁻³ mol NaOH

EXAMINER'S TOP TIP
You need the equation here.

Stage 3: Finding the concentration of hydrochloric acid.

$$\text{concentration} = \frac{\text{no. of mol}}{\text{volume of acid (dm}^3)}$$

$$= \frac{2.50 \times 10^{-3}}{19.80 \times 10^{-3}}$$

$$= 0.127 \text{ mol dm}^{-3}$$

Labels on diagram: pipette filler, burette, burette clamp, pipette, acid, jet, tap, jet, conical flask, alkali and indicator, white tile

Acid–base titrations can be used to measure the concentration of an acid solution.

Quick test

1 You have to make 250 cm³ of sodium carbonate solution, Na_2CO_3 of concentration 0.02 mol dm⁻³. Describe how you would do this.

2 Titration can be used to find the volume of an acid solution needed to react with 25.0 cm³ of a base if you have two standard solutions.

Calculate the volume of hydrochloric acid (0.10 mol dm⁻³) needed to react with 25.0 cm³ of sodium carbonate solution (0.22 mol dm⁻³).

The equation for the reaction is

$$Na_2CO_3(aq) + 2HCl(aq) \rightarrow 2NaCl(aq) + CO_2(g) + H_2O(g)$$

3 Sodium carbonate reacts with hydrochloric acid with a different indicator. The indicator changes colour when half the volume of acid used in Q2 is added. Write an equation for this reaction.

Use data from the Data section (page 91) to help you answer these questions.

1 Balance the following equations.
  a $Mn_3O_4 + - Al \rightarrow - Mn + - Al_2O_3$
  b $NH_3 + - O_2 \rightarrow - NO + - H_2O$
  c $Na_2H_2P_2O_7 + - NaHCO_3 \rightarrow - Na_2HPO_4 + - CO_2 + - H_2O$ [3]

2 What is the minimum volume of hydrochloric acid (4.0 mol dm^{-3}) required to dissolve 0.1 mole of magnesium according to the following equation?
  $Mg(s) + 2H^+(aq) \rightarrow Mg^{2+}(aq) + H_2(g)$ [3]

3 A mixture of magnesium chloride and magnesium sulphate is known to contain 0.6 moles of chloride ions and 0.2 moles of sulphate ions. What is the number of magnesium ions present?

  ....

  ....

  ....   [3]

4 Calculate the concentration of sulphuric acid (in mol dm^{-3}) if 25.0 cm^3 of 0.10 mol dm^{-3} sodium hydroxide solution is neutralised by 20.0 cm^3 of sulphuric acid.
  $2NaOH(aq) + H_2SO_4(aq) \rightarrow Na_2SO_4(aq) + 2H_2O(l)$

  ....

  ....

  ....

  ....   [4]

5 In a power station, sulphur dioxide is removed from waste gases by passing it through an aqueous suspension of calcium carbonate. The calcium sulphate produced is used to make plasterboard.
  The equation for the reaction is:
  $2SO_2(g) + 2CaCO_3(s) + O_2(g) \rightarrow 2CaSO_4(s) + 2CO_2(g)$
  a What is the mass of 2 moles of calcium sulphate?

  ....   [1]

  b How many tonnes of calcium carbonate are needed to produce 1 tonne of calcium sulphate?

  ....   [1]

  c How many tonnes of sulphur dioxide are needed to produce 1 tonne of calcium sulphate?

  ....   [1]

  d If 1 mole of sulphur dioxide has a volume of 24 dm^3 at r.t.p., what volume of sulphur dioxide produces 1 tonne of calcium sulphate?

  ....

  ....

  ....   [3]

6 Write chemical equations for the following reactions.

  a Solid sodium carbonate is heated with calcium hydroxide solution to give a precipitate of calcium carbonate and sodium hydroxide solution.

  .... [3]

  b Magnesium metal displaces silver from silver nitrate solution. Magnesium nitrate solution is also formed.

  .... [3]

  c Solid copper(II) nitrate decomposes on heating to form copper(II) oxide, nitrogen dioxide and oxygen.

  .... [3]

  d Ethane(C_2H_6) burns in excess oxygen to form carbon dioxide and water.

  .... [3]

7 An element X forms compounds XCl_2 and XBr_2. The dibromide is completely converted into the dichloride when it is heated in a stream of chlorine.

  $$XBr_2 + Cl_2 \rightarrow XCl_2 + Br_2$$

  When 1.500 g of XBr_2 is treated, 0.890 g of XCl_2 is formed.
  Calculate the relative atomic mass of X and, using the Periodic Table, identify the element X.

  ....

  ....

  ....

  .... [3]

8 $Na_2CO_3(aq) + 2HCl(aq) \rightarrow 2NaCl(aq) + CO_2(g) + H_2O(l)$
  1.40 g of pure anhydrous sodium carbonate (Na_2CO_3, $M_r=106$) were dissolved in water and made up to a volume of 250 cm^3.
  A 25.0 cm^3 sample of this solution required 24.5 cm^3 of hydrochloric acid to neutralise it.
  Calculate the concentration of this hydrochloric acid in g dm^{-3}.

  ....

  ....

  ....

  .... [4]

9 2.32 g of a pure sample of an iron oxide contains 1.68 g of iron.
  a Calculate the formula of this iron oxide.

  [4]

  b This iron oxide is reduced to iron by hydrogen. Write a balanced equation for this reaction.

  .... [3]

Total: /45

Atomic structure

- Atoms are the smallest part of an element that can take part in a chemical reaction.
- Until just over 100 years ago it was believed that atoms were indivisible, like snooker balls.
- We now know that atoms are made up of three types of particle: <u>protons</u>, <u>neutrons</u> and <u>electrons</u>.

Particles in an atom

The table compares the properties of these three particles.

Particle	Mass	Charge
proton, p	1 amu	+1
neutron, n	1 amu	0
electron, e	negligible	−1

(amu = atomic mass unit)

An **atom** is neutral, i.e. it contains an equal number of protons and electrons.

If an atom loses or gains one or more electrons it becomes a charged **ion**.

Proton number (or atomic number) is the number of protons (or electrons) in an atom.

Mass number is the number of protons plus neutrons in an atom.

Atoms containing the same number of protons but different numbers of neutrons are called **isotopes**. e.g. Chlorine has a proton number of 17. It exists as two isotopes: chlorine-35 and chlorine-37. Chlorine-35 contains 17p, 17e and 18n; chlorine-37 contains 17p, 17e and 20n.

Structure of the atom

The protons and neutrons in an atom are tightly packed together in the positively charged **nucleus**. The electrons are around the nucleus in certain **energy levels**.
Each energy level can only hold up to a certain number of electrons.
The table summarises the maximum number of electrons in each shell.

Energy level or shell	Maximum number of electrons
1st shell	2
2nd shell	8
3rd shell	18
4th shell	32
5th shell	50

The diagram shows a simple representation of a sodium atom. The electron arrangement of sodium can be summarised as 2,8,1.

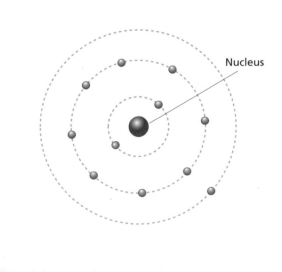

Nucleus

Advanced model of electron arrangement

Spectroscopic studies have showed that the eight electrons in the 2nd shell are not identical in energy. The shells can be broken down into **subshells**.

Electrons, because of their very small size, are impossible to locate exactly at a particular time. It is possible to show a region or volume where the electron is likely to be. This is called an **orbital**. Orbitals can be divided into s, p, d and f types. The diagrams show the shapes of s and p orbitals.

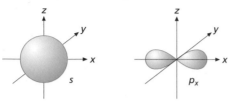
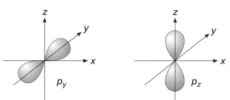

The first energy level holds a maximum number of two electrons in one s type orbital (called 1s). there are no p, d or f orbitals available at this energy level. The second energy level consists of one s type orbital and three types of p orbitals: $2s$, $2p_x$, $2p_y$, $2p_z$.

Note: there are three p orbitals of identical energy – one along the x-axis, one along the y-axis and one along the z-axis.

These four orbitals can hold a total of eight electrons (i.e. two electrons each). There are no $2d$ or $2f$ orbitals.

The third energy level consists of one s type orbital, three p type orbitals and five d type orbitals. These nine orbitals can hold a maximum of 18 electrons altogether (two electrons each). You are not expected to know the shapes of d or f orbitals.

When filling the available orbitals with electrons two important principles should be remembered:

1. Electrons fill the lowest energy orbitals first and other orbitals in order of ascending energy. This is called the **Aufbau Principle**. The order of filling orbitals is $1s$, $2s$, $2p$, $3s$, $3p$, $4s$, $3d$, $4p$, $5s$, $4d$, $5p$, $6s$, $4f$, $5d$, $6p$, etc. This is summarised in the diagram.

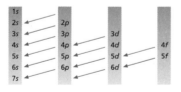

2. Where there are several orbitals of exactly the same energy (e.g. three $2p$ orbitals), electrons will occupy different orbitals whenever possible (e.g. nitrogen is $1s^2 2s^2 2p_x^1 2p_y^1 2p_z^1$ and not $1s^2 2s^2 2p_x^2 2p_y^1$). This is called **Hund's Rule**. An orbital containing only one electron has the electron **unpaired**.

The table shows the electron arrangement of the first 20 elements in a simple form suitable for GCSE and a form suitable for AS or A2.

Element	GCSE level	A level
Hydrogen	1	$1s^1$
Helium	2	$1s^2$
Lithium	2, 1	$1s^2 2s^1$
Beryllium	2, 2	$1s^2 2s^2$
Boron	2, 3	$1s^2 2s^2 2p_x^1$
Carbon	2, 4	$1s^2 2s^2 2p_x^1 2p_y^1$
Nitrogen	2, 5	$1s^2 2s^2 2p_x^1 2p_y^1 2p_z^1$
Oxygen	2, 6	$1s^2 2s^2 2p_x^2 2p_y^1 2p_z^1$
Fluorine	2, 7	$1s^2 2s^2 2p_x^2 2p_y^2 2p_z^1$
Neon	2, 8	$1s^2 2s^2 2p_x^2 2p_y^2 2p_z^2$
Sodium	2, 8, 1	$1s^2 2s^2 2p_x^2 2p_y^2 2p_z^2 3s^1$
Magnesium	2, 8, 2	$1s^2 2s^2 2p_x^2 2p_y^2 2p_z^2 3s^2$
Aluminium	2, 8, 3	$1s^2 2s^2 2p_x^2 2p_y^2 2p_z^2 3s^2 3p_x^1$
Silicon	2, 8, 4	$1s^2 2s^2 2p_x^2 2p_y^2 2p_z^2 3s^2 3p_x^1 3p_y^1$
Phosphorus	2, 8, 5	$1s^2 2s^2 2p_x^2 2p_y^2 2p_z^2 3s^2 3p_x^1 3p_y^1 3p_z^1$
Sulphur	2, 8, 6	$1s^2 2s^2 2p_x^2 2p_y^2 2p_z^2 3s^2 3p_x^2 3p_y^1 3p_z^1$
Chlorine	2, 8, 7	$1s^2 2s^2 2p_x^2 2p_y^2 2p_z^2 3s^2 3p_x^2 3p_y^2 3p_z^1$
Argon	2, 8, 8	$1s^2 2s^2 2p_x^2 2p_y^2 2p_z^2 3s^2 3p_x^2 3p_y^2 3p_z^2$
Potassium	2, 8, 8, 1	$1s^2 2s^2 2p_x^2 2p_y^2 2p_z^2 3s^2 3p_x^2 3p_y^2 3p_z^2 4s^1$
Calcium	2, 8, 8, 2	$1s^2 2s^2 2p_x^2 2p_y^2 2p_z^2 3s^2 3p_x^2 3p_y^2 3p_z^2 4s^2$

Note: sometimes electronic structures are shown in a slightly condensed form, e.g. calcium $1s^2 2s^2 2p^6 3s^2 4s^2$

Quick test

1 A fluorine atom has a proton number of 9 and a mass number of 19. How many protons, neutrons and electrons are there in an atom of fluorine?

2 Without looking in the table, write down the full electron arrangement of lithium (proton number 3) and oxygen (proton number 8).

3 Which orbitals are filled when 18 electrons are in an atom?

1. 9 protons, 10 neutrons and 9 electrons. 2. Lithium $1s^2 2s^1$; oxygen $1s^2 2s^2 2p_x^2 2p_y^1 2p_z^1$ 3. $1s2s 2p,3s 2p_y,2p_x 2p, 3s, 3p_x, 3p_y, 3p_z$

23

Mass spectrometer

- **Atoms of different elements have different masses.**
- **The masses of individual atoms are too small to be weighed.**
- **A mass spectrometer is an instrument that can be used to compare the masses of different atoms.**

Structure of the mass spectrometer

The diagram shows the main parts of a mass spectrometer.

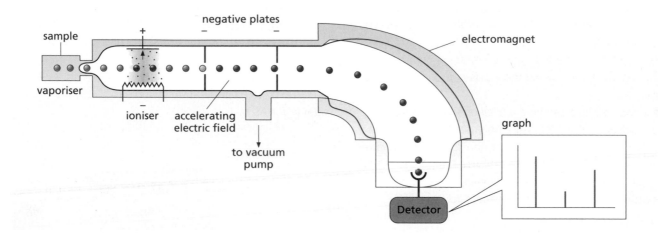

The table shows the functions of the main parts of a mass spectrometer.

Part of mass spectrometer	Function
Vaporiser	The sample is heated and turned to a gas.
Ioniser	Atoms are converted to positive ions after collisions with fast moving electrons.
Accelerating magnetic field	Speeds up the positive ions.
Vacuum pump	Removes air from the apparatus. Otherwise oxygen and nitrogen could be ionised and ions could collide with particles in the air.
Electromagnet	The magnetic field causes the ions to change direction. The field is constantly changing.
Detector	This detects the ions and the results are displayed on a graph.

Principle of the mass spectrometer

The sample is ionised and the ions are accelerated and passed through a constantly changing magnetic field. The ions are deflected.

The deflection of an ion depends upon its mass/charge ratio.

The mass spectrometer in action

The sample, if not gaseous, is vaporised.

It passes at very low pressure into the main chamber.

The vapour is bombarded with high-energy electrons and some collisions will result in electron-loss from the sample with the formation of positive ions.

The ions are accelerated by passage through slits in two negatively charged plates. At this stage, the positive ions have similar velocities but different charge/mass ratios.

The ions pass through a magnetic field and are deflected by it (the deflection produced by a given magnetic field depending on the mass/charge (m/z) ratio of the ion).

The magnetic field is gradually adjusted so that ions with different charge/mass ratios are focused, in turn, on the detector.

The mass of an ion and its relative abundance are indicated by a peak on a chart.

The mass spectrum of chlorine gas

The diagram shows the mass spectrum of chlorine gas.

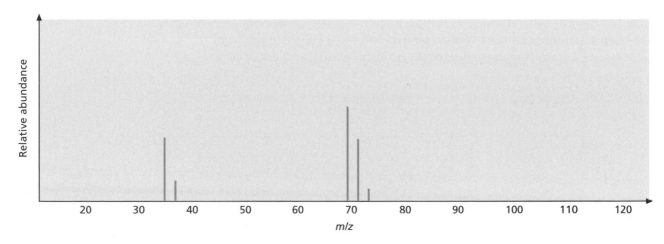

Chlorine consists of two isotopes: chlorine-35 and chlorine-37.
Chlorine molecules could be $^{35}Cl–^{35}Cl$ or $^{37}Cl–^{37}Cl$ or $^{35}Cl–^{37}Cl$
and free chlorine atoms either ^{35}Cl or ^{37}Cl.
Ionisation produces $Cl_2 \rightarrow 2Cl^+ + 2e^-$ or $Cl_2 \rightarrow Cl_2^+ + e^-$
Peak at 74 corresponds to Cl_2^+ where both chlorines are chlorine-37.
Peak at 72 corresponds to Cl_2^+ where one chlorine is chlorine-37 and one chlorine-35.
Peak at 70 corresponds to Cl_2^+ where both chlorines are chlorine-35.
Peak at 37 corresponds to Cl^+ where chlorine is chlorine-37.
Peak at 35 corresponds to Cl^+ where chlorine is chlorine-35.

Quick test

1 **Write equations for the ionisation of hydrogen gas.**

2 **Helium gas can be ionised to form different ions. Write down equations for the formation of He^+ or He^{2+}. Which ion will be deflected more.**

3 **Why is there a large peak at 15 and a small peak at 17?**

4 **The diagram shows the spectrum for methane CH_4**

   a **Explain why the relative molecular mass of methane is 16.**

   b **Why are there large peaks at 15 and 16 but a small peak at 17?**

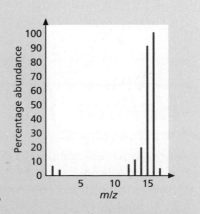

Ionic bonding

- **Ionic bonding involves the complete transfer of one or more electrons from a metal atom to a non-metal atom.**
- **Positive and negative ions are formed.**
- **Ions are held together by strong electrostatic forces.**
- **Compounds with ionic bonding are usually high melting point solids that dissolve in water to form solutions that conduct electricity.**

Lewis structures

The American chemist Gilbert Lewis in 1916 drew simple structures of atoms of elements showing only the electrons in the outer shell. These diagrams are called **Lewis Structures**.
The outer electrons, called **valence electrons**, are the ones that are involved in bonding.

The diagrams show the Lewis structures of the first twenty elements.

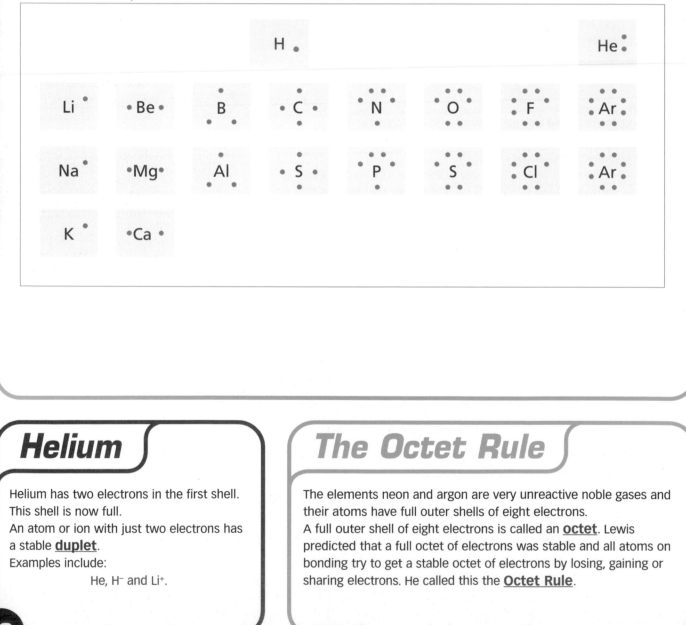

Helium

Helium has two electrons in the first shell. This shell is now full.
An atom or ion with just two electrons has a stable **duplet**.
Examples include:

He, H⁻ and Li⁺.

The Octet Rule

The elements neon and argon are very unreactive noble gases and their atoms have full outer shells of eight electrons.
A full outer shell of eight electrons is called an **octet**. Lewis predicted that a full octet of electrons was stable and all atoms on bonding try to get a stable octet of electrons by losing, gaining or sharing electrons. He called this the **Octet Rule**.

Sodium chloride

Atoms can achieve noble gas electron arrangement by loss or gain of electrons to form ions. Metals (with low electronegativities) lose electrons to form positive ions and non-metals gain electrons to form negative ions.

The most common example of ionic bonding is sodium chloride. A complete transfer of one electron from a sodium atom to a chlorine atom leads to the formation of a positive sodium ion and a negative chloride ion. These ions are held together by strong electrostatic forces. Both sodium ions and chloride ions have full octets. The changes are summarised by the Lewis structure.

The changes in atomic arrangement are summarised by

Magnesium oxide

Another example of ionic bonding is magnesium oxide. In this case each magnesium atom loses two electrons to form an ion with a 2+ charge. Each oxygen atom gains two electrons to form an ion with a 2– charge. The changes are summarised by the Lewis structure.

The changes in atomic arrangement are summarised by

The properties of compounds with ionic bonding are discussed later in the 'Structure of solids' section.

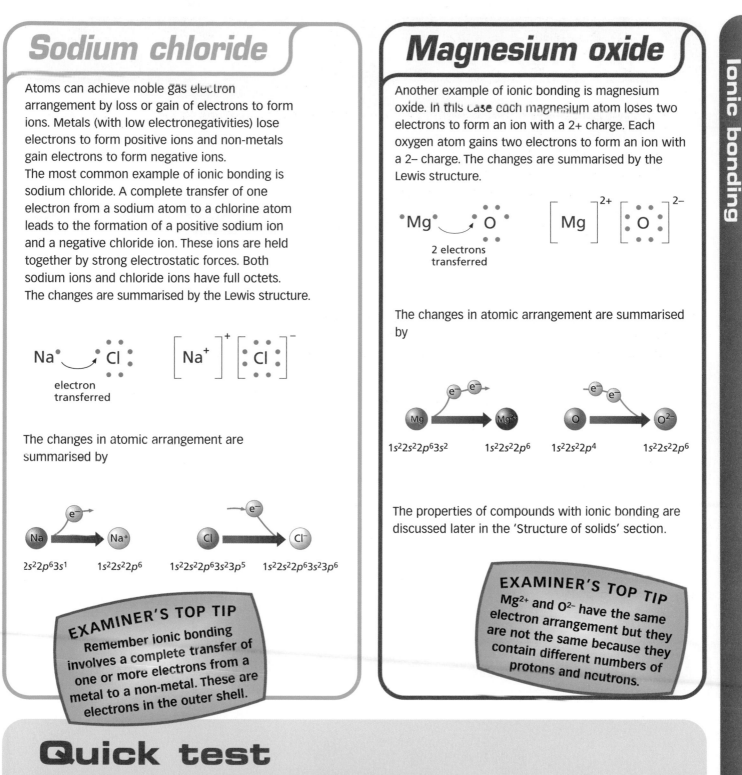

$2s^22p^63s^1$ $\quad$ $1s^22s^22p^6$ $\quad$ $1s^22s^22p^63s^23p^5$ $\quad$ $1s^22s^22p^63s^23p^6$

$1s^22s^22p^63s^2$ $\quad$ $1s^22s^22p^6$ $\quad$ $1s^22s^22p^4$ $\quad$ $1s^22s^22p^6$

EXAMINER'S TOP TIP
Remember ionic bonding involves a complete transfer of one or more electrons from a metal to a non-metal. These are electrons in the outer shell.

EXAMINER'S TOP TIP
Mg^{2+} and O^{2-} have the same electron arrangement but they are not the same because they contain different numbers of protons and neutrons.

Quick test

1 *Look back at the Lewis structures of lithium and fluorine.*

   a *What happens to the lithium atom when lithium and fluorine bond?*

   b *What happens to the fluorine atom when lithium and fluorine bond?*

   c *Write down the formulae of the ions formed.*

2 *The formula of sodium oxide is Na_2O. What changes take place to the sodium and oxygen atoms when they combine?*

3 *Sodium hydride is an ionic compound. Describe the changes that take place when sodium and hydrogen combine.*

4 *Aluminium fluoride is formed when aluminium and fluorine combine. Describe the changes in electron arrangement that take place.*

1. (a) The lithium atom loses an electron; (b) The fluorine atom gains an electron; (c) Li^+ and F^-. 2. Each sodium atom loses one electron forming a Na^+ ion. The oxygen atom gains two electrons. 3. Each sodium atom loses one electron forming a Na^+ ion. Each hydrogen atom gains one electron forming an H^- ion. 4. Each aluminium atom loses 3 electrons and forms Al^{3+} ions. Each fluorine atom gains 1 electron and forms F^- ions.

Covalent bonding

- Covalent bonding is bonding usually between two non-metal atoms.
- A covalent bond involves the sharing of a pair of electrons, with one electron coming from each atom.
- In some cases two pairs of electrons are shared between two atoms and this is called a double bond.
- Covalent bonding was first proposed by Gilbert Lewis in 1916. When two atoms combine by covalent bonding they <u>share</u> a pair of electrons with one electron coming from each atom.

Hydrogen molecule

The simplest case of covalent bonding is within a hydrogen molecule. Each hydrogen atom has a single electron in a 1s orbital. It is highly unlikely that one hydrogen atom will lose an electron and the other atom gain an electron to produce ions. Instead, each atom gives its electron to form a **shared pair** of electrons in a **molecular orbital** between the two hydrogen atoms. This is called a **covalent bond** and each hydrogen atom has a share of two electrons. The bonding in a hydrogen molecule can be shown in the simplified diagrams below.

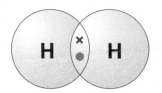

The small dots and crosses represent valence electrons (i.e. electrons in the outer shell) and the single stroke represents a covalent bond.

Other molecules with covalent bonding

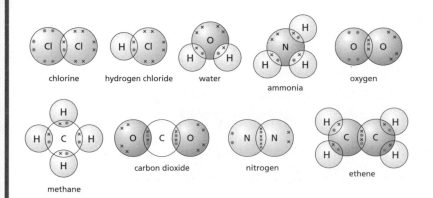

chlorine   hydrogen chloride   water   ammonia   oxygen

methane   carbon dioxide   nitrogen   ethene

In all the above examples covalent bonding leads to the formation of small molecules. It is possible for covalent bonding to produce an extremely large arrangement of atoms called a **giant structure**. Silicon(IV) oxide (silicon dioxide) is an example of this and can be represented in a simplified form in the diagram.

- silicon
- oxygen

structure continues indefinitely

all bonds are strong bonds

Molecules containing covalent bonding

Substances made up of molecules containing covalent compounds exhibit the following features.

- They have low melting and boiling points (they are usually gases or low boiling point liquids at room temperature).
- They are usually soluble in organic solvents such as hexane.
- They are usually insoluble in water. However, hydrogen chloride is very soluble in water because it ionises into H^+ and Cl^- ions in water.

Coordinate (or dative covalent) bonding

Coordinate bonding is a type of covalent bonding.
A pair of electrons is shared between two atoms to form a bond but one atom supplies both electrons and the other atom supplies none.
An example of coordinate bonding is the compound formed between boron trifluoride BF_3 and ammonia NH_3. Both electrons in the N and B bond are donated by the nitrogen atom. The molecule produced is shown by the Lewis structure below.

There are many examples of coordinate bonding with transition metal ions.
For example:

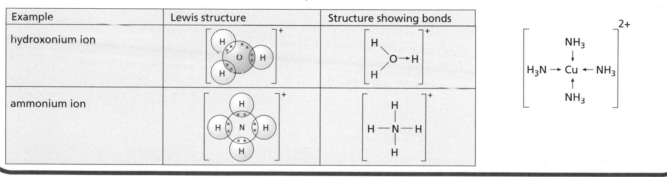

Example	Lewis structure	Structure showing bonds

Quick test

1 Ethane has a formula C_2H_6.

  a Draw a Lewis structure of an ethane molecule.

  b Is this a solid, liquid or gas at room temperature?

2 Draw a Lewis structure of a fluorine molecule, F_2.

3 Ethyne has a formula C_2H_2. Draw a Lewis structure of a molecule of ethyne.

4 The diagram shows a molecule of aluminium chloride.

  a The aluminium chloride is an electron deficient molecule. Explain this statement.

  b Aluminium chloride is found to have a formula Al_2Cl_6. Draw a diagram showing how coordinate bonding leads to this.

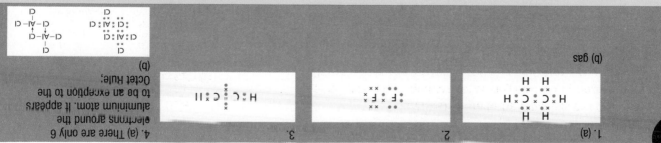

Shapes of molecules

- The shapes of molecules are determined by the number of electron pairs there are around the central atom.
- The pairs of electrons can either be bonded pairs or lone pairs.
- Lone pairs have a larger repulsive effect than bonded pairs and distort the shape of the molecule more.

Electron-pair repulsion theory

There are two important points:

1. The shape of a molecule or ion is determined by the number of electron pairs in the outer shell of the central atom.
2. The electron pairs repel one another and get as far apart as possible.

Molecules with bonded pairs

Number of electron pairs	Example	Dot and cross diagram	Shape	Bond angles
2	Beryllium chloride	$:Cl \overset{x}{\underset{x}{:}} Be \overset{x}{\underset{x}{:}} Cl:$	Linear Cl—Be—Cl 180°	180°
3	Boron trifluoride	F, B, F, F	Trigonal planar F—B 120° F F	120°
4	Methane	$H \overset{x}{\underset{x}{:}} C \overset{x}{\underset{x}{:}} H$ ... H	Tetrahedral H—C 109.5° H H H	109.5°
5	Phosphorus pentachloride	Cl Cl P Cl Cl Cl	Trigonal tripyramid Cl, Cl—P—Cl 90° 120° Cl Cl	120° and 90°
6	Sulphur hexafluoride	F F S F F F F	Octahedral F F—S—F 90° F F F	90°

Molecules with lone pairs

Ed: can only enlarge the a/w so much.
It still doesn't fill all the space.

Two common examples are water and ammonia.

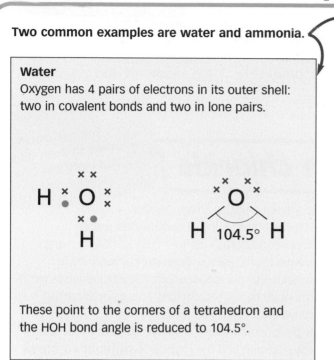

Water
Oxygen has 4 pairs of electrons in its outer shell: two in covalent bonds and two in lone pairs.

These point to the corners of a tetrahedron and the HOH bond angle is reduced to 104.5°.

Ammonia
Nitrogen has 3 pairs of electrons in its outer shell: three in covalent bonds and one in a lone pair.

These point to the corners of a tetrahedron and the HNH bond angles are reduced to 107°.

Molecules with double bonds

In molecules where there is a double bond, the double bond is treated in the same way as a bonded pair.

Carbon dioxide, CO_2

180°

O = C = O

Sulphur dioxide, SO_2

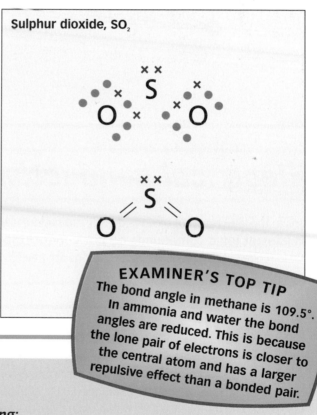

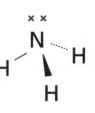

EXAMINER'S TOP TIP
The bond angle in methane is 109.5°. In ammonia and water the bond angles are reduced. This is because the lone pair of electrons is closer to the central atom and has a larger repulsive effect than a bonded pair.

Quick test

1 *Predict the shape and bond angle of the following:*
  a *BeF_2*
  b *H_2S*
  c *PF_3*
  d *CS_2*
  e *$SiCl_4$*

2 *Explain why the bond angle in ammonia is less than the bond angle in methane.*

1. (a) linear 180°; (b) non linear 104.5°; (c) pyramidal 107°; (d) linear 180°; (e) tetrahedral 109.5° 2. The repulsion between lone pair and bonded pair is greater than the repulsion between two bonded pairs.

Structure of solids

There are four types of crystal structure:

- **ionic**, e.g. sodium chloride
- **metallic**, e.g. gold
- **molecular**, e.g. iodine
- **giant covalent**, e.g. diamond and graphite.

Ionic structure of sodium chloride

Compounds such as sodium chloride and magnesium oxide have large numbers of positive and negative ions packed together in a regular arrangement, called a **lattice**. This diagram shows the lattice of sodium chloride.

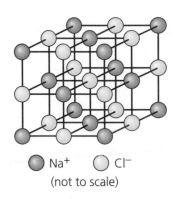

● Na⁺   ○ Cl⁻
(not to scale)

The sodium chloride lattice consists of a face-centred cube of sodium ions and an interpenetrating face centred cube of chloride ions. In simple terms, each sodium ion in the lattice is surrounded by six chloride ions and each chloride ion is surrounded by six sodium ions. This **coordination** here is called 6:6.

The electrostatic forces between the ions in the lattice are very strong and the **lattice is difficult to break up**. As a result compounds with ionic bonding are **solids with high melting points**. The lattice can be broken up by melting or by dissolving in water. Compounds containing ionic bonding **do not dissolve in organic solvents** such as hexane.

Electrical conductivity

Ions in the lattice cannot move and there are no free electrons.
As a **result ionic compounds do not conduct electricity when solid**.
When melted or in aqueous solution the ions are free to move and so **electricity is conducted in melts and solutions**.

Metallic structures

A metal consists of a close-packed, regular arrangement of positive metal ions, which are surrounded by a 'sea' of delocalised electrons. The ions within each close-packed layer are arranged in a regular hexagon.

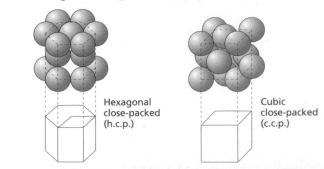

Hexagonal close-packed (h.c.p.)

Cubic close-packed (c.c.p.)

The close-packed layers of ions can be stacked in two ways – both being equally likely. The two ways of stacking are:

- ABAB or **hexagonal close-packing**. The ions in the third layer are immediately above ions in the first layer. Magnesium and zinc have this structure.
- ABC or **cubic close-packing** (sometimes called **face-centred cubic**). The ions in the third layer are not immediately above ions in the first layer.

In both structures each ion in the structure is surrounded by twelve other ions (the **coordination number** is 12): six in the same layer, three in the layer above and three in the layer below.

Molecular structure of iodine

Iodine is an example of a substance with covalent crystals. Within the crystals the iodine molecules are regularly arranged. The I–I bonds are strong but the forces between the molecules are weak. The crystal structure breaks down when heated to a low temperature.

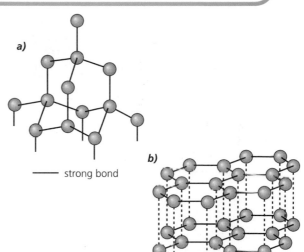

32°

Iodine crystal

The structures of diamond and graphite

Diamond and graphite are two **giant covalent structures (macromolecular)**. Both have a regular arrangement of carbon atoms.

In the diamond structure the carbon atoms are each bonded to four other carbon atoms. All bonds are very strong.

Fullerenes

Until about 20 years ago, chemists believed that diamond and graphite were the only two crystalline forms of carbon. Then Harry Kroto and his team at the University of Sussex discovered a new form of carbon. This new form consists of clusters of carbon atoms, either C_{60} or C_{70}. These are formed when carbon is evaporated from carbon electrodes in helium gas at low pressure.

a)

—— strong bond

b)

------ weak bond

In graphite the carbon atoms are arranged in layers. In each layer carbon atoms are strongly bonded in hexagonal rings. There are only very weak forces between the layers. The layers are able to slide over one another. In a graphite structure there are delocalised electrons that enable electricity to be conducted through the structure.

Quick test

1 Write down three properties of a substance containing ionic bonding.

2 Explain why magnesium oxide, MgO, has a much higher melting point than sodium chloride, NaCl.

3 Carbon dioxide is a gas but silicon dioxide is a solid. Explain this difference.

4 Explain the properties of diamond and graphite. Use the structures to help you.

1. High melting point; soluble in water; conduct electricity when molten or in aqueous solution.   2. In magnesium oxide the ions are Mg^{2+} and O^{2-}. The electrostatic forces between ions with 2+ and 2− charges is much greater so a higher temperature is needed to melt the solid.   3. Carbon dioxide is made up of small molecules. The bonds within each molecule are covalent. There are no forces between the molecules. In silicon(IV) oxide all atoms are bonded together into a giant structure.   4. Diamond is very hard. The structure is only broken down when strong bonds are broken. Diamond has no delocalised electrons and so does not conduct electricity. Graphite is very soft as layers can slide over each other without breaking strong bonds. Graphite conducts electricity because it has delocalised electrons.

Intermolecular forces

- The forces within molecules are <u>strong covalent bonds</u>.
- There are weak forces between molecules called <u>intermolecular forces</u>.
- These forces are <u>Van der Waals' forces</u>, <u>permanent dipole–dipole attractions</u> and <u>hydrogen bonds</u>.

Van der Waals' Forces

Van der Waals' forces (or induced dipole–dipole interactions) are:
- very weak intermolecular forces

The small dipole is produced by a temporary uneven distribution of electrons within the atom of molecule. When two atoms or molecules approach one another, slight charges are momentarily induced due to slight movement of electrons.

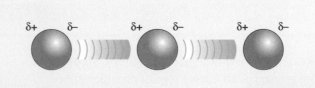

movement of electrons produces an oscillating dipole

oscillating dipole induces dipole in next atom

Van der Waals' forces increase in strength with increasing numbers of electrons.
Van der Waals' forces are found in alkanes (page 70) and noble gases.

Noble gas	Boiling point/°C	Number of electrons	Trend
helium	−269	2	As Group is descended, atoms get larger with more electrons. Electron clouds more easily distorted so induced dipoles increase and Van der Waals' forces are stronger. Boiling points increase.
neon	−246	10	
argon	−186	18	
krypton	−152	36	
xenon	−107	54	
radon	−62	86	

Permanent dipole–dipole attractions

Small dipoles exist in polar molecules due to movement of electrons in covalent bonds.
The small + and − ions cause attraction between adjacent molecules.
An example is in propanone.

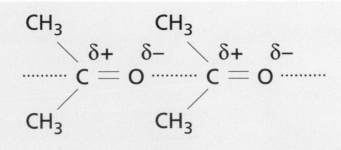

Hydrogen bonding

Hydrogen bonding is more commonly encountered between molecules (**intermolecular**) rather than within molecules (**intramolecular**).

It occurs in compounds whose molecules consist of a hydrogen atom covalently bonded to an electronegative atom: usually fluorine, oxygen or nitrogen.

There are slight charges within the molecules caused by a slight shift of electrons in the covalent bond. As a result of these charges, weak electrostatic attractions exist between molecules and these are called **hydrogen bonds**.

For example, water:

$$\delta+ \quad \delta- \qquad \delta+ \quad \delta- \qquad \delta+ \quad \delta-$$
$$H-\ddot{\underset{..}{F}}: \qquad H-\ddot{\underset{..}{O}}: \qquad H-N<$$

This association of molecules has no effect on chemical properties but alters physical properties. For example, the boiling point is increased as extra energy is required to break these bonds before the molecules can escape from the liquid and boiling takes place. These extra bonds also increase the viscosity and surface tension of the liquid.

The boiling point of water is much higher than would be expected by comparison with similar compounds (see page 5).

Other examples of hydrogen bonding

Organic acids such as ethanoic acid exist as dimers when in nonpolar solvents such as methylbenzene.

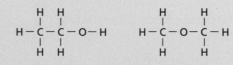

EXAMINER'S TOP TIP

It is easy to confuse these different types of intermolecular forces and not to appreciate the relative strength of these forces. Ionic and covalent bonds are strong bonds. A hydrogen bond is about one tenth of the strength of a covalent bond. A permanent dipole–dipole bond is much weaker and van der Waals' forces are much weaker than permanent dipole–dipole bonds.

Quick test

1 **Ethanol and methoxymethane have the same molecular mass but ethanol has a much higher boiling point.**

   Explain the different boiling points.

2 **Which of the following have hydrogen bonding?**

   H_2 HCl HF F_2 CH_3F CH_3OH

3 **Butan-1-ol and butanal have approximately the same molecular mass.**

   Why does butan-1-ol have a higher boiling point (117 °C) than butanal (80 °C)?

1. Hydrogen bonding exists in ethanol because hydrogen atom is attached to an electronegative oxygen. Hydrogen bonding cannot exist in methoxymethane, only permanent dipole–dipole and van der Waal's forces that are much weaker.   2. HF and CH₃OH (remember for hydrogen bonding a hydrogen atom must be covalently bonded to an electronegative atom; e.g. F, O or N).   3. Butan-1-ol has hydrogen bonding but butanal does not.

35

Enthalpy changes – 1

- Many chemical reactions are accompanied by energy changes.
- In some cases energy is lost by the reacting chemicals to the surroundings.
- In other cases energy is gained by the reacting chemicals from the surroundings.

Enthalpy

Enthalpy, H, is the **heat energy** that is stored within a chemical system.
It is impossible to measure the enthalpy in a system by experiment.
It is possible to measure the **enthalpy change**, ΔH, from temperature changes.
The enthalpy change is the enthalpy exchange between the system and the surrounding at constant pressure.

Exothermic reactions

A reaction where heat energy is **released** to the surroundings is called an exothermic reaction.
The same quantity of energy is lost by the reacting chemicals as is gained by the surroundings.
It causes the temperature of the surroundings to rise.
An exothermic reaction can be summarised by a **reaction profile**.
Notice that the value for **ΔH is negative**.

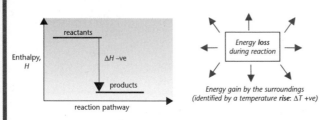

Energy is required to break chemical bonds and is released when new bonds are formed.
In an exothermic reaction, more energy is released from the forming of new bonds than is required to break existing bonds.

There are many examples of exothermic reactions including combustion and rusting of iron.

Endothermic reactions

In an endothermic reaction heat energy is **taken in** from the surroundings.
The gain in energy by the reacting chemicals is matched by a loss in energy by the surroundings. This causes the temperature of the surroundings to fall.
An endothermic reaction can be summarised by a **reaction profile**.
Notice that the value for **ΔH is positive**.

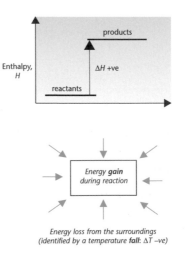

Energy loss from the surroundings
*(identified by a temperature **fall**: ΔT –ve)*

In an endothermic reaction, less energy is released from the forming of new bonds than is required to break existing bonds.

There are few examples of endothermic reactions. As chemists could not find examples of endothermic reactions, they once proposed that only exothermic reactions could take place.

Exothermic and endothermic reactions

Here are a couple of examples.

1. Hydrochloric acid mixed with sodium hydroxide solution is an exothermic reaction. If solutions of hydrochloric acid and sodium hydroxide, both at 20 °C, are mixed, the temperature of the solution rises.

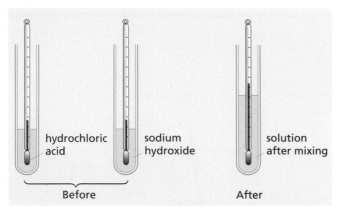

Before / After

Energy released during the reaction to the surroundings causes the temperature to rise.

2. Citric acid mixed with sodium hydrogencarbonate solution produces an endothermic reaction. If crystals of citric acid are added to sodium hydrogencarbonate solution at 20 °C, the temperature of the solution falls.

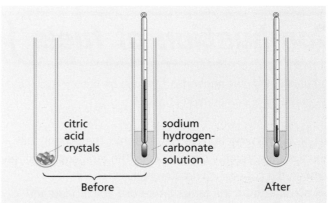

citric acid crystals / sodium hydrogen-carbonate solution

Before / After

Energy is taken in by the reacting chemicals causing the temperature of the surroundings to fall.

Respiration and photosynthesis

Respiration is an **exothermic reaction**.
In the body the energy released maintains body temperature and provides the energy the body needs.

$$C_6H_{12}O_6 + 6O_2 \rightarrow 6CO_2 + 6H_2O + energy$$

Photosynthesis is an **endothermic process**. This process enables green plants to make food.

$$6CO_2 + 6H_2O + energy \rightarrow C_6H_{12}O_6 + 6O_2$$

Quick test

1 Draw diagrams to show the energy profiles for respiration and photosynthesis.

2 Explain why the same temperature rise recorded when hydrochloric acid (2 mol dm^{-3}) is added to either sodium hydroxide (2 mol dm^{-3}) or potassium hydroxide (2 mol dm^{-3}) is the same.

3 The equation for the reaction of hydrogen and chlorine is:

$H_2(g) + Cl_2(g) \rightarrow 2HCl(g)$

   a Draw an energy profile for this exothermic reaction.

   b Explain what happens in the reaction in terms of bond breaking and bond making.

   c The bond energies for H–H, Cl–Cl and H–Cl are 435, 243 and 431 kJ mol^{-1}, respectively. Calculate the value of ΔH.

(b) H–H and Cl–Cl bonds are broken (this requires energy). H–Cl bonds are formed (this releases energy); (c) $(435 + 243) - (431 \times 2) = -184$ kJ)

1. 2. In both cases the same reaction is taking place – $H^+(aq) + OH^-(aq) \rightarrow H_2O(l)$ 3. (a) $H_2(g) + Cl_2(g)$

Enthalpy changes – 2

- Enthalpy changes can be measured by experiment.
- They can also be calculated using bond enthalpy data (see Data section on pages 90-91)

Combustion of fuels

The diagram shows apparatus that can be used to calculate the enthalpy of combustion of a fuel.

All combustion reactions are exothermic. The burning fuel releases energy to the surroundings. The temperature of the water in the copper calorimeter rises.
It is possible, from the temperature rise, to calculate the amount of energy gained by the water in the calorimeter. By weighing the spirit lamp before and after it is possible to calculate the mass of fuel burned.

Then it is possible to calculate the energy released when 1 mole of fuel is completely burned. This is the **enthalpy of combustion**.

e.g. for the fuel propanone, CH_3COCH_3:

Temperature of water before	= 20.0 °C
Temperature after	= 42.0 °C
Temperature rise	= 22.0 °C
Mass of propanone burned during experiment	
	= 1.16 g
Mass of water in calorimeter	= 250 g

Ignore the heat capacity of the copper calorimeter and the thermometer.

Energy transferred to water (J)
= mass of water (g) × temperature rise (K)
× specific heat capacity of water (J⁻¹ mol K⁻¹)

Energy transferred	= 250 × 22.0 × 4.2
	= 23 100 J
	= 23.1 kJ
The mass of 1 mole of propanone	= (12 × 3) + (6 × 1) + 16
	= 58 g

1.16 g of propanone produces 23.1 kJ

58 g of propanone produces $\dfrac{23.1}{1.16} \times 58 = 1155$ kJ mol⁻¹

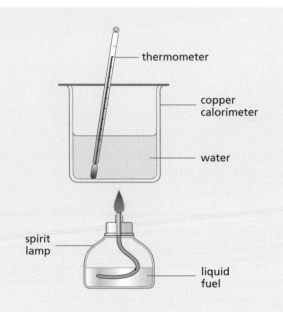

The data book value for the enthalpy of combustion of propanone is 1786 kJ mol⁻¹.
The difference between the experimental value and the data book value is very different. This is because not all the energy released when the propanone burns heats up the water.
Some will be used to heat up the air and surrounding equipment.

EXAMINER'S TOP TIP
The specific heat capacity of a material is the energy needed to raise the temperature of 1 g by 1K. For water and dilute solutions, the specific heat capacity is 4.2 Jg⁻¹K⁻¹.

Hess's Law

The energy change for a reaction is the same whether a reaction takes place in one stage or in a series of stages. This can be summarised by the diagram:

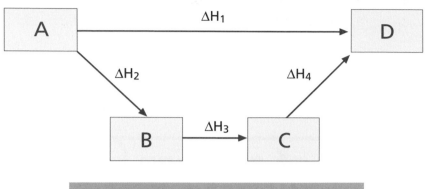

$\Delta H_1 = \Delta H_2 + \Delta H_3 + \Delta H_4$
If three of these values are known the fourth one can be calculated.

Quick test

1 The diagram shows apparatus that can be used to calculate the enthalpy of neutralisation in the reaction of sodium hydroxide and hydrochloric acid.

When 50 cm³ of 2.0 mol dm⁻³ hydrochloric acid was mixed with 50 cm³ of 2.0 mol dm⁻³ sodium hydroxide solution the temperature rise was 13.3 °C.

a What is the advantage of using a polystyrene cup rather than a metal beaker?

b Calculate the energy gained by the solution when the two solutions are mixed.

c Calculate the numbers of moles of hydrochloric acid and sodium hydroxide used.

d Calculate the enthalpy of neutralisation i.e. the energy when 1 mole of acid and base react.

2 The equation summarises the reaction of ethene and hydrogen to form ethane.

Use the data on bond enthalpies (pages 90-91) to calculate the ΔH for this reaction.

1. (a) Polystyrene is a poor conductor of heat and so energy losses from the polystyrene cup are less than from a metal beaker;
(b) energy gained by solution = mass × temperature rise × specific heat = 100 × 13.3 × 4.2 = 5586 J (This makes the approximation that 50 cm³ of hydrochloric acid and 50 cm³ of sodium hydroxide solution have a mass of 100 g and specific heat capacity of the solution is 4.2 Jg⁻¹K⁻¹); (c) numbers of moles of hydrochloric acid

$$= \frac{50 \text{ dm}^3 \times 2.0 \text{ mol dm}^{-3}}{1000}$$

= 0.1 mol

Number of moles of sodium hydroxide = 0.1 mol;

(d) enthalpy of neutralisation $= -\frac{5586 \text{ J}}{1000} = -55.9$ kJ mol⁻¹

N.B. There is a negative sign because the reaction is exothermic.

2. Bonds broken		
	C=C	+612 kJ
	H–H	+436 kJ
		+1048 kJ

Bonds formed		
	C–C	–348 kJ
	2 C–H	–824 kJ
		–1142 kJ

ΔH = +1048 kJ – 1172 kJ = –124 kJ

Reaction kinetics

● The factors that can alter the rate of a reaction are:

concentration

temperature

physical state of reactants, e.g. lumps, powder

catalyst

light

EXAMINER'S TOP TIP
You should have studied the effect of these factors for GCSE. You need to go into more detail now.

Measuring rate of reaction

In a reaction A → B the rate of reaction can be measured as a decrease in the concentration of A in a given time or an increase in the concentration of B in a given time. The units of rate of reaction are mol dm^{-3} s^{-1}.

The graph summarises the changes in concentration of A, [A], and concentration of B, [B], with time.

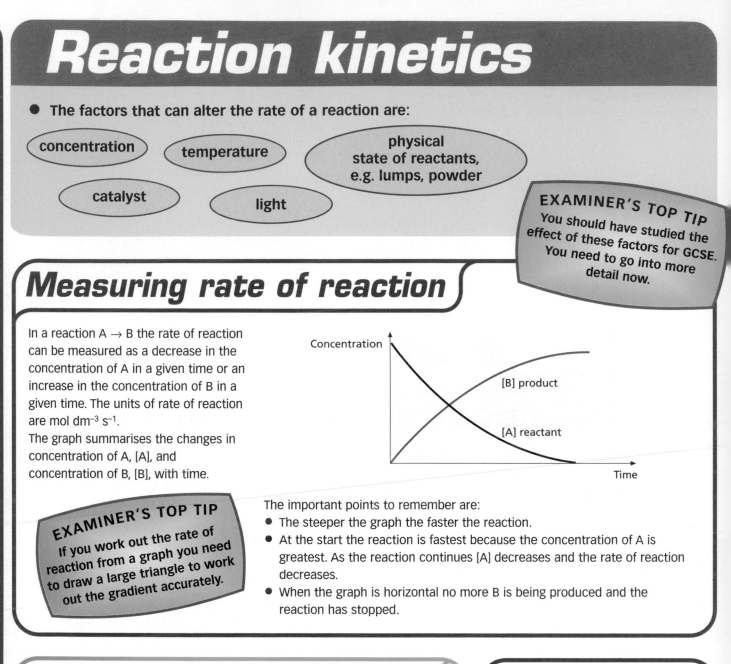

EXAMINER'S TOP TIP
If you work out the rate of reaction from a graph you need to draw a large triangle to work out the gradient accurately.

The important points to remember are:

● The steeper the graph the faster the reaction.
● At the start the reaction is fastest because the concentration of A is greatest. As the reaction continues [A] decreases and the rate of reaction decreases.
● When the graph is horizontal no more B is being produced and the reaction has stopped.

Physical state of reactants

Reactions occur much faster when reactants are in a powdered form. This provides a larger surface area for reaction to take place.
For example, the reaction of calcium carbonate and hydrochloric acid.

$$CaCO_3(s) + 2HCl(aq) \rightarrow CaCl_2(aq) + H_2O(l) + CO_2(g)$$

The graph shows the results of an experiment with lumps and powdered calcium carbonate.

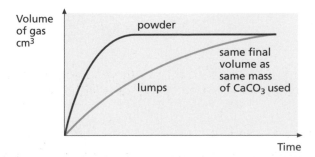

The reaction is faster with a powder than with lumps as the powder has a much larger surface area. There will be more collisions per second between the surface of the powder and the acid.

Temperature

The diagram shows an energy level diagram. Before a reaction can take place some energy, called the **activation energy**, E_a, has to be provided to start the reaction.

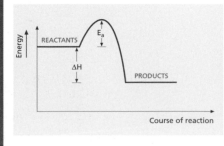

Boltzmann Distribution

The graph, called the Boltzmann Distribution, shows the distribution of molecular energies of the particles in a gas at constant temperature.

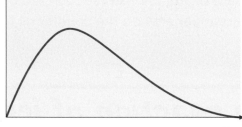

The average kinetic energy of the particles in a gas increases as temperature increases. The next graph shows the distribution of particles of different energies at two temperatures T_1 and T_2.

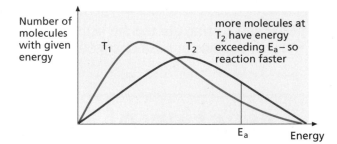

more molecules at T_2 have energy exceeding E_a – so reaction faster

Remember:
- There are the same number of particles in each case.
- At a higher temperature the peak of the curve is moved to the right and the curve is flatter.
- More particles exceed the activation energy, E_a, leading to a faster reaction.

Catalysts

A catalyst is a substance which alters the rate of a chemical reaction without being used up.

e.g. $2 H_2O_2 \rightarrow 2H_2O + O_2$ Catalyst: manganese(IV) oxide

A catalyst does not alter the position of an equilibrium or increase the yield of products. It merely alters the rate at which an equilibrium is reached or the products are obtained.

It provides an alternative route with a lower activation energy barrier for the reaction. More molecules possess the lower activation energy and so the reaction is speeded up.

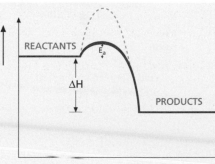

There are different types of catalyst:
- **Heterogeneous catalyst** – Solid where reaction occurs on the surface, e.g. iron in the Haber process to manufacture ammonia.
- **Homogeneous catalyst** – Catalyst completely mixes with the reactants. Acid acts as a catalyst in the formation of an ester.

Light

Some reactions are affected by light, e.g. the reaction between hydrogen and chlorine (page 85) is faster than light.

EXAMINER'S TOP TIP
Remember homo- means same and hetro- means different. Homogeneous catalyst is in the same phase as reactants. Heterogeneous catalyst is in a different phase to reactants.

Quick test

1 The reaction of calcium carbonate with dilute hydrochloric acid can be followed by measuring the volume of gas produced at intervals. Suggest an alternative way of doing this.

2 Explain, using ideas of particles, why:

a Increasing temperature increases the rate of reaction.

b Crushing up limestone into a powder before adding to acid, increases the rate of reaction.

1. Carbon dioxide escapes from the reaction mixture. The reaction can be followed by measuring the loss of mass at intervals. 2. (a) Increasing temperature increases the average kinetic energy of the particles. When they collide more of them have sufficient energy to exceed the activation energy and so the reaction is faster; (b) Crushing the powder gives a larger surface area. There can be more collisions between the acid particles and the limestone and so the reaction is faster.

Reversible reactions and equilibrium

- **Reactions that can go both in the forward direction and the reverse reaction are called <u>reversible reactions</u>.**
- **If a reversible reaction takes place in a closed system an equilibrium can be set up.**
- **In an equilibrium, the rate of the forward and the rate of the reverse reaction are equal.**
- **Le Chatalier's Principle can be used to predict what happens when the conditions in a system in equilibrium are changed.**

The reaction of iron and steam

When steam is passed over heated iron, a reaction takes place and iron oxide and hydrogen are formed.

$$3Fe(s) + 4H_2O(g) \rightleftharpoons Fe_3O_4(s) + 4H_2(g)$$

The hydrogen is collected in a test tube over water and excess steam condenses.

The reaction of iron oxide and hydrogen

When hydrogen is passed over heated iron oxide the iron oxide is reduced to iron. Steam is also produced and condenses in the cooled U tube.

$$Fe_3O_4(s) + 4H_2(g) \rightleftharpoons 3Fe(s) + 4H_2O(g)$$

The unreacted hydrogen escapes from the apparatus.

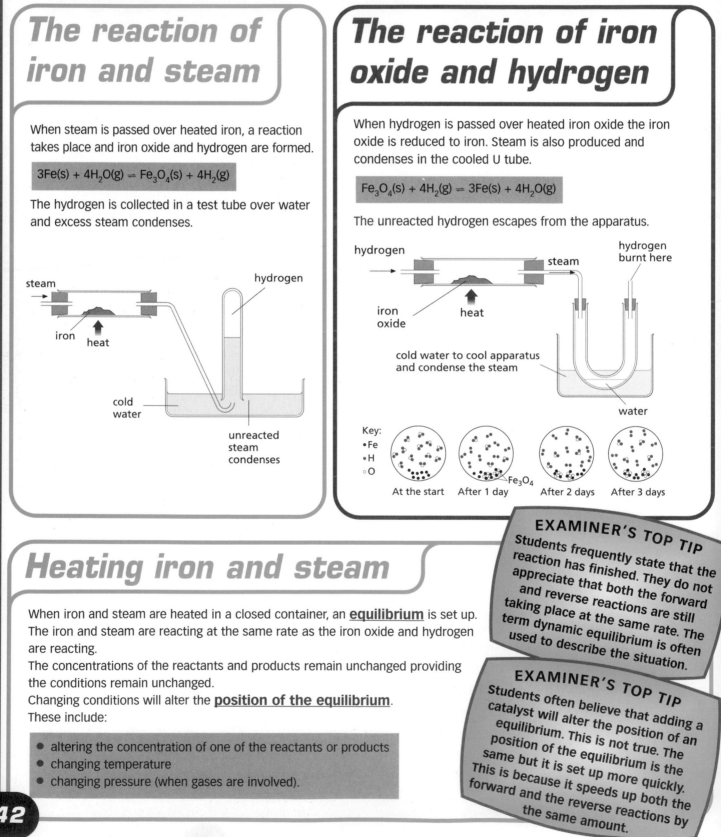

Heating iron and steam

When iron and steam are heated in a closed container, an **equilibrium** is set up. The iron and steam are reacting at the same rate as the iron oxide and hydrogen are reacting.

The concentrations of the reactants and products remain unchanged providing the conditions remain unchanged.

Changing conditions will alter the **position of the equilibrium**. These include:

- altering the concentration of one of the reactants or products
- changing temperature
- changing pressure (when gases are involved).

EXAMINER'S TOP TIP
Students frequently state that the reaction has finished. They do not appreciate that both the forward and reverse reactions are still taking place at the same rate. The term dynamic equilibrium is often used to describe the situation.

EXAMINER'S TOP TIP
Students often believe that adding a catalyst will alter the position of an equilibrium. This is not true. The position of the equilibrium is the same but it is set up more quickly. This is because it speeds up both the forward and the reverse reactions by the same amount.

Moving an equilibrium

The equilibrium **moves to the right** when [A] and [B] decrease, and [C] and [D] increase.

$$A + B \rightleftharpoons C + D$$

The equilibrium **moves to the left** when [A] and [B] increase, and [C] and [D] decrease.

EXAMINER'S TOP TIP
Remember – decreasing [A] and [B] will have the opposite effect as increasing [A] and [B] and the same effect as increasing [C] and [D].

Le Chatalier's Principle

Le Chatalier's Principle states that:
if a system in equilibrium is subjected to a change, the system will change to minimise this change and to restore equilibrium.

Altering the concentration of a reactant or product

$$A + B \rightleftharpoons C + D$$

Change	Effect of the change
Increasing [A] or [B]	Increases the rate of the forward reaction. As more products are formed rate of reverse reaction increases. New equilibrium set up with smaller [C] and [D].
Increasing [C] or [D]	Increases the rate of the reverse reaction. As more reactants are formed rate of forward reaction increases. New equilibrium set up with smaller [A] and [B].

Effect of temperature changes

The effect on the equilibrium depends upon whether the reaction is **exothermic** or **endothermic**.

Reaction is exothermic, i.e. ΔH is negative	Effect on equilibrium
Increasing temperature	Equilibrium moves to the the left, so more reactants and less products
Decreasing temperature	Equilibrium moves to the right, so less reactants and more products.

If the reaction is endothermic, i.e. ΔH is positive, the effect is opposite.

The effect of pressure

The position of the equilibrium will be changed only if:
- At least one of the products or reactants is a gas.
- There are different numbers of gaseous moles of reactants and products.
  e.g. $2SO_2(g) + O_2(g) \rightleftharpoons 2SO_3(g)$

This equation fits the two conditions above so altering pressure alters the position of equilibrium.

The position of equilibrium moves towards the direction where there is the fewer number of gas molecules.

In this case increasing pressure moves equilibrium to the right (fewer molecules).

The synthesis of ammonia is an example of a reaction where altering conditions alters the position of the equilibrium (see page 63).

Quick test

What happens to the position of the equilibrium in each of the following examples when:
(a) temperature is increased, and (b) pressure increased?

1 $N_2O_4(g) \rightleftharpoons 2NO_2(g)$, endothermic.

2 $2SO_2(g) + O_2(g) \rightleftharpoons 2SO_3(g)$, exothermic.

3 $CO_2(g) + H_2(g) \rightleftharpoons CO(g) + H_2O(g)$, endothermic.

4 Why is the position of the equilibrium not changed when pressure is increased in the reaction below?

$$CH_3COOH(l) + C_2H_5OH(l) \rightleftharpoons CH_3COOC_2H_5(l) + H_2O(l)$$

1. Equilibrium moved: (a) to the right; (b) to the left 2. Equilibrium is changed: (a) to the left; (b) to the right 3. Equilibrium is not changed by increasing pressure as there are the same number of molecules on both sides. Equilibrium would be moved to the right by increasing the temperature. 4. Equilibrium is unchanged as there are no gases involved in the reaction.

Acids and bases

- Acids are compounds containing hydrogen atoms which release hydrogen ions when added to water.

- Acids are compounds containing hydrogen

- Bases are metal oxides and hydroxides. Bases that are soluble in water release the hydroxide ion and are called alkalis.

Common acids and bases

Acids such as hydrochloric acid, HCl, sulphuric acid, H_2SO_4, and nitric acid, HNO_3, are called **mineral acids**. Ethanoic acid and citric acid are examples of other acids.

Common bases include:
- metal oxides, e.g. calcium oxide, CaO, and copper(II) oxide, CuO
- metal hydroxides, e.g. calcium hydroxide, $Ca(OH)_2$, and sodium hydroxide, NaOH
- ammonia and amines, e.g. ammonia, NH_3, and methylamine, CH_3NH_2.

Common properties of acids

Although acids vary considerably, they do have some similar properties.

1 Reaction with metals, e.g. magnesium.
 Magnesium reacts with an acid to produce hydrogen gas.
 e.g. $Mg(s) + 2HCl(aq) \rightarrow MgCl_2(aq) + H_2(g)$
 $Mg(s) + H_2SO_4(aq) \rightarrow MgSO_4(aq) + H_2(g)$

EXAMINER'S TOP TIP
You should be able to write ionic equations for these reactions.
e.g. $Mg(s) + 2H^+(aq)$
$\rightleftharpoons Mg^{2+}(aq) + H_2(g)$

2 Reaction with a carbonate or hydrogencarbonate.
 A carbonate or hydrogencarbonate react with an acid to produce carbon dioxide.
 e.g. $Na_2CO_3(s) + 2HCl(aq) \rightarrow 2NaCl(aq) + CO_2(g) + H_2O(l)$
 $NaHCO_3(s) + HCl(aq) \rightarrow NaCl(aq) + CO_2(g) + H_2O(l)$

3 Reaction with a base.
 An acid reacts with a base to form a salt and water only. Copper(II) oxide is often used.
 e.g. $CuO(s) + 2HCl(aq) \rightarrow CuCl_2(aq) + H_2O(l)$
 $CuO(s) + H_2SO_4(aq) \rightarrow CuSO_4(aq) + H_2O(l)$

Strong and weak acids

Acids can be classified as strong acids and weak acids.
A **strong acid** is an acid which is completely dissociated into ions in solution.
A **weak acid** is only partially dissociated into ions. For example:

hydrochloric acid $HCl(g) + H_2O(l) \rightarrow H_3O^+(aq) + Cl^-(aq)$
hydrofluoric acid $HF(g) + H_2O(l) \rightleftharpoons H_3O^+(aq) + F^-(aq)$

The H_3O^+ ion is called the **oxonium** ion. It is sometimes written as $H^+(aq)$

Do not confuse strength of an acid with concentration. An acid's concentration gives the number of particles of acid in a given volume of solution. The strength of an acid refers to the amount of dissociation in the acid. Also, do not confuse strength with the danger of using an acid. Hydrofluoric acid is a weak acid but is very dangerous to use.

Strong and weak alkalis

Some bases dissolve in water to form **alkalis**. Alkalis contain the **hydroxide** ion, OH^-.
Alkalis can be divided into **strong and weak alkalis**.
A strong alkali is completely dissociated into ions and a weak alkali is only partially dissociated into ions.

> For a **strong alkali**, e.g. sodium hydroxide:
> $$NaOH(s) + aq \rightarrow Na^+(aq) + OH^-(aq)$$

> For a **weak alkali**, e.g. ammonia solution:
> $$NH_3(g) + H_2O(l) \rightleftharpoons NH_4^+(aq) + OH^-(aq)$$

Brønsted–Lowry Theory of acids and bases

A Brønsted–Lowry acid is a **proton donor**.

A Brønsted–Lowry base is a **proton acceptor**.

According to the definition of Brønsted–Lowry, when hydrochloric acid dissociates in water it donates a proton to the water to form the oxonium ion. The water molecule is acting as a base as it is accepting a proton.
When ammonia dissolves in water, ammonia accepts a proton to form an ammonium ion. Ammonia is therefore a base.

Conjugate acids and bases

The equation for the reaction that occurs when ammonia is added to water is shown below.

$$NH_3(g) + H_2O(l) \rightleftharpoons NH_4^+(aq) + OH^-(aq)$$

In the forward reaction, ammonia is acting as an acid (donating protons).
In the reverse reaction, ammonium ion acts as an acid (proton donor) and hydroxide ion acts as a base (proton acceptor).

In any acid–base equilibrium there are two acids and two bases. An acid turns into its **conjugate base** and a base turns into its **conjugate acid**. In this example, the two conjugate acid–base pairs are:

H_2O and OH^-   **NH_4^+ and NH_3**

Quick test

1 Sulphuric acid is a diprotic acid. Dissociation of both hydrogen atoms occurs.
   Write the equation for the dissociation of sulphuric acid in water.

2 Ethanoic acid, CH_3COOH, dissociates partially into ethanoate and hydrogen ions.
   Write the equation for the dissociation of ethanoic acid in water.

   Is ethanoic acid a strong acid or a weak acid?

3 Explain why a solution of hydrochloric acid (0.1 mol dm^{-3}) has a lower pH than a
   solution of ethanoic acid (0.1 mol dm^{-3}).

4 An equilibrium is set up when methanoic acid is added to water.

   $$HCOOH(l) + H_2O(l) \rightleftharpoons HCOO^-(aq) + H_3O^+(aq)$$

   What are the two conjugate acid-base pairs?

1. $H_2SO_4(l) + 2H_2O(l) \rightarrow 2H_3O^+(aq) + SO_4^{2-}(aq)$ 2. $CH_3COOH(l) + H_2O(l) \rightleftharpoons H_3O^+(aq) + CH_3COO^-(aq)$. It is a weak acid.
3. The solutions have the same concentration but the hydrochloric acid is completely dissociated and the ethanoic acid is only partially dissociated. There are far more oxonium ions in hydrochloric acid.   4. acid–base: methanoic acid–methanoate ion; oxonium ion–water.

Physical chemistry

Use data from the Data section (pages 90–91) to help you answer these questions.

1 How many protons, neutrons and electrons are there in each of the following atoms?

a ^{137}Ba .... [1]

b ^{235}U .... [1]

c ^{238}U .... [1]

2 The element gallium contains 60.2% of an isotope of mass number 69 and 39.8% of an isotope of mass number 71. Calculate the relative atomic mass of gallium. [4]

3 There are three isotopes of hydrogen. Explain the similarities and differences in the structure of these three isotopes. [3]

....

....

....

4 Write down the full electron arrangements of the following elements in terms of *s,p* and *d* orbitals.

a phosphorus .... [1]

b oxygen .... [1]

c titanium .... [1]

d bromine .... [1]

e arsenic .... [1]

5 A sample of oxygen consisting mainly of the isotope-16 was enriched with oxygen-18. The composition of the mixture was 73.0% oxygen-16 and 25.0% oxygen-18 by volume.

The oxygen sample reacted with sodium as follows:

$4Na(s) + O_2(g) \rightarrow 2Na_2O(s)$

a Finish the table.

Species	protons	neutrons	electrons
sodium-23 atom			
oxygen-16 atom			
oxide-18 ion			

[9]

b Write down the full electron arrangement of:
(i) a sodium atom [1] **(ii)** an oxide ion. [1]

(i) ....

(ii) ....

c State three properties of sodium oxide. [3]

....

....

....

d Some carbon-12 was burned in another sample of the oxygen mixture. The carbon dioxide gave the following mass spectrum:

Identify the peaks shown. [3]

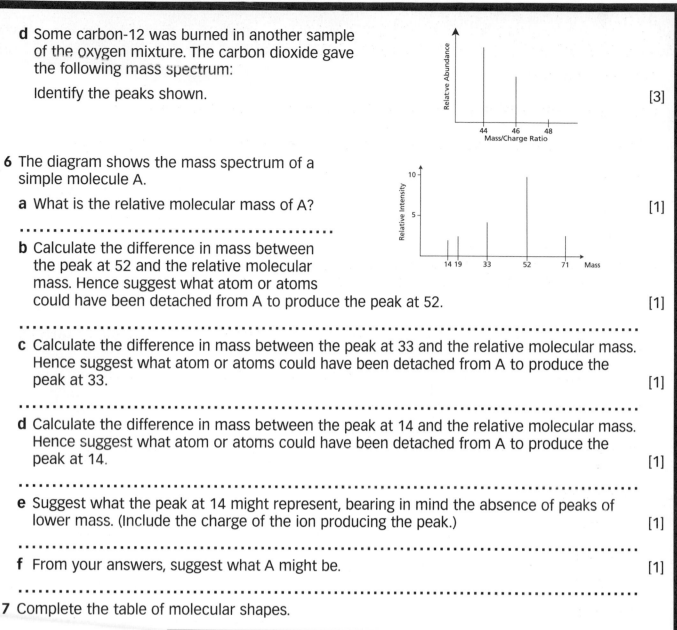

6 The diagram shows the mass spectrum of a simple molecule A.

a What is the relative molecular mass of A? [1]

....

b Calculate the difference in mass between the peak at 52 and the relative molecular mass. Hence suggest what atom or atoms could have been detached from A to produce the peak at 52. [1]

....

c Calculate the difference in mass between the peak at 33 and the relative molecular mass. Hence suggest what atom or atoms could have been detached from A to produce the peak at 33. [1]

....

d Calculate the difference in mass between the peak at 14 and the relative molecular mass. Hence suggest what atom or atoms could have been detached from A to produce the peak at 14. [1]

....

e Suggest what the peak at 14 might represent, bearing in mind the absence of peaks of lower mass. (Include the charge of the ion producing the peak.) [1]

....

f From your answers, suggest what A might be. [1]

....

7 Complete the table of molecular shapes.

Number of pairs of bonding electrons	Number of lone pairs of electrons	Shape of molecule
3	0	
3	1	
4	0	
2	2	

[4]

8 Using the bond enthalpies given on page 91, calculate the enthalpy change for the following reaction. [3]

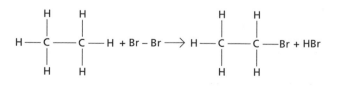

Total: / 45

47

The Periodic Table

The Periodic Table is an arrangement of all the <u>chemical elements</u> in order of increasing <u>atomic number</u> – elements with similar properties (i.e. the same chemical family) are placed in the same <u>vertical column</u>.

You will find understanding the Periodic Table will enable you to make <u>predictions</u> about elements and their <u>properties</u>.

You will find a full copy of the Periodic Table on page 90.

The first Periodic Table

The Periodic Table was devised in 1869 by Mendeleef. At this time a number of elements that we now know had not been discovered. This made the task more difficult. Mendeleef left gaps in his table and even predicted the properties of some of these undiscovered elements.

The main difference between Mendeleef's table and the modern table, apart from the extra elements, is that the latter arranges the elements in order of increasing atomic number rather than atomic mass.

Structure of the Periodic Table

The diagram shows the main parts of the modern Periodic Table.

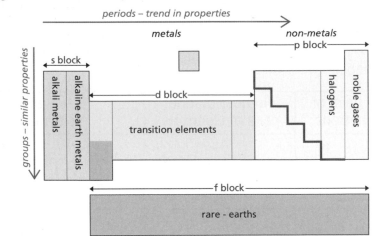

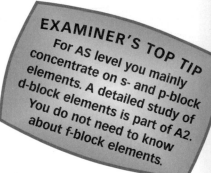

EXAMINER'S TOP TIP
For AS level you mainly concentrate on s- and p-block elements. A detailed study of d-block elements is part of A2. You do not need to know about f-block elements.

- The vertical columns are called **groups**. A group contains elements with similar properties and similar outer electron arrangements. The groups are sometimes given Roman numbers, e.g. Group IV. The number of electrons in the outer shell is equal to the group number, e.g. the elements in the alkali metal family (Group 1) all have single electrons in the outer shell. (Noble gases in Group 0 are an exception.)

- The horizontal rows in the Periodic Table are called **periods**. An element in Period 3 has 3 shells of electrons.

- The bold line in the Periodic Table is an attempt to divide elements into metals and non-metals. Elements on the right-hand side of this line are non-metals and on the left-hand side are metals. Elements close to the line often show both metallic and non metallic properties, particularly if they exist in allotropic (polymorphic) forms.

- The elements shaded in pink are the **s-block elements** and the ones shaded in yellow are the **p-block elements**. The s- and p-block elements make up the **main block of elements**.

- The elements between the s and p blocks are called **d-block** or **transition elements** (shaded in green).

- The two rows of elements at the bottom of the table (shaded in purple) are called the **f-block elements**.

Periodicity

Physical properties such as ionisation energy, electronegativity, electron affinity, melting and boiling points are related to electron arrangements. If one of these properties is plotted on a graph for each element the graph consists of a series of peaks and troughs. Such a graph is said to be **periodic**.

Ionisation energy

The graph shows the first ionisation energy of each element plotted against atomic number.

Look at the graph.
The elements at the top of each peak are noble gases (from Group 0).
The elements in the bottom of each trough are alkali metals (Group 1).

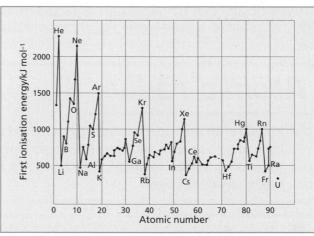

Electronegativity

The tendency of an atom to attract electrons to itself in a covalent bond is called electronegativity.
The graph shows the electronegativity of each element plotted against atomic number.

Again a periodic graph is obtained.

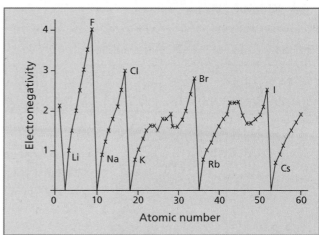

Quick test

1 Look at the Periodic Table on page 90. Classify each of these elements as s-block, p-block, d-block or f-block.

   a silicon   d strontium

   b cobalt   e iodine

   c uranium

2 Classify each of these elements as metals or non-metals.

   a bromine   d selenium

   b calcium   e manganese

   c chromium

3 Look at the graph of electronegativity against atomic number.

   a Elements from which group are at the top of each peak?

   b Elements from which group are at the bottom of each trough?

   c How does the electronegativity change across each period?

   d How does the electronegativity change down each group?

3. (a) group 7 (Halogens); (b) group 0 (Noble gases); (c) increases left to right; (d) decreases down group.
1. (a) p-block; (b) d-block; (c) f-block; (d) s-block; (e) p-block   2. (a) non-metal; (b) metal; (c) metal; (d) non-metal; (e) metal

49

Atomic and ionic radii

There are trends in the sizes of atoms and ions both across a period of the periodic table and also down a group.

These trends can be explained in terms of <u>numbers of shells or orbitals or electrons</u>, <u>nuclear charge</u> and the <u>screening by electrons</u> in inner shells.

Atomic radius

The <u>orbitals</u> of an atom extend to <u>infinity</u> and so there is <u>no definite size</u> to the atom.
In order to assign a size, it is necessary to make some arbitrary decision as to where the <u>boundary</u> of an atom is.
One way is to say that the <u>covalent radius</u> is half the internuclear distance between two atoms of an element joined by a <u>single covalent bond</u>.

Trend across a period

The atomic radii of the elements in the second period of the Periodic Table are:

	Li	Be	B	C	N	O	F
atomic radius / nm	0.123	0.106	0.088	0.077	0.070	0.066	0.064

EXAMINER'S TOP TIP
Many students fail to realise that across a period there is a decrease in atomic radius. It seems unlikely since with each move to the right, one proton and one electron have been added, and possibly neutrons as well. You must remember that there is a lot of space within the atom.

The trends across a period are:

- The nuclear charge increases. Lithium contains 3 protons and beryllium contains 4 protons.
- The outer electrons are being added to the same shell. In moving across this period all electrons are going in the 3rd shell (into $3s$ or $3p$ orbitals).
- The attraction between the nucleus and the outer electrons increases.

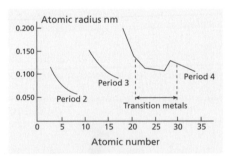

As a result <u>the radius of the atom decreases gradually across the period</u>. The same is true in other periods.

Trend down a group

The table shows the simple electron arrangement of atoms in Group 7 and simple diagrams showing the electron arrangement in these atoms.

Element	Electron arrangement	Simple diagrams of electron arrangement	Atomic radius/nm
Fluorine	2, 7		0.064
Chlorine	2, 8, 7		0.099
Bromine	2, 8, 18, 7		0.111
Iodine	2, 8, 18, 18, 7		0.128

Trend down a group continued

The trends down a group are:

- There are extra shells that are further from the nucleus.
- As the outer electrons become further from the nucleus the attraction between the other electrons and the nucleus decreases.
- Electrons in the inner shell repel electrons in the outer shell. This is called **shielding**.

The atomic radius increases down a group.

Atomic radii (in nm) and relative sizes

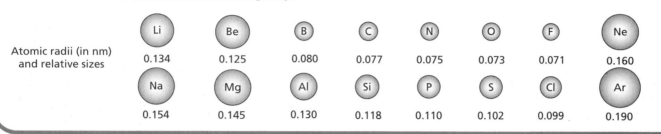

Li	Be	B	C	N	O	F	Ne
0.134	0.125	0.080	0.077	0.075	0.073	0.071	0.160
Na	Mg	Al	Si	P	S	Cl	Ar
0.154	0.145	0.130	0.118	0.110	0.102	0.099	0.190

Ionic radius

The trends of ionic radii down a group are similar to those for atomic radii.

Cations (or positive ions)

They have a smaller radius than the corresponding atoms,

e.g. **Na 0.157 nm, Na$^+$ 0.098 nm.**

The removal of one electron from a sodium atom empties the $3s$ orbital. The outer electron is therefore in a $2p$ orbital.

Anions (or negative ions)

They have a larger radius than the corresponding atoms because of repulsion between the electrons. There is no extra nuclear charge.

e.g. **F 0.064 nm, F$^-$ 0.133 nm**

Isoelectronic ions

The three ions Na$^+$, Mg^{2+} and Al^{3+} contain the same number of electrons in an arrangement of $1s^2 2s^2 2p^6$. They are said to be **isoelectronic**.

The ionic radii of these ions are 0.098, 0.065 and 0.045 nm respectively.

The decreasing radius is due to the additional nuclear charge: Na$^+$ contains 11 protons, Mg^{2+} has 12 and Al^{3+} has 13.

Ion	Ionic radius/nm	Ion	Ionic radius/nm
Li$^+$	0.068	F$^-$	0.133
Na$^+$	0.098	Cl$^-$	0.181
K$^+$	0.133	Br$^-$	0.196
Rb$^+$	0.148	I$^-$	0.219
Cs$^+$	0.167		

Quick test

1 Arrange the following in order of increasing size.

K   K$^+$ Rb

Explain your answer.

2 The elements in period 3 are:

Na   Mg   Al   Si   P   S   Cl   Ar

a What trend would be seen in atomic radius from left to right?

b Explain why this trend exists.

3 Write down the symbol for an atom and an ion which are isoelectronic with Cl$^-$.

and the outer electrons.   3. Ar and S^{2-} (or P^{3-}).

nuclear charge but the extra electrons go into the same orbitals 3s and 3p (in the 3rd shell). Increasing attraction between the nucleus

electron is lost. A rubidium atom has 5 shells of electrons.   2. The atomic radius decreases from left to right. There is increasing

1. K$^+$ < K < Rb. A potassium atom has 4 shells of electrons with a single electron in the 4s orbital. When the K$^+$ ion is formed this outer

51

Oxidation states

The system of oxidation states (or oxidation numbers) is used as a guide to the extent of oxidation or reduction in a species.

The system is useful because it can be used for both ionic and covalent bonding.

Oxidation state

The oxidation state is simply the number of electrons which must be added to a positive ion to get a neutral atom or removed from a negative ion to get a neutral atom, e.g.

> $Fe^{2+}(aq)$: two electrons have to be added so the oxidation state is +2;
> $Cl^-(aq)$: one electron has to be removed therefore the oxidation state is –1.

For covalent species it is assumed that the electrons in the covalent bond actually go to the atom which is most electronegative, e.g.

> ammonia, NH_3, nitrogen is the more electronegative element. It is assumed that the three electrons (one from each hydrogen atom) are associated with the nitrogen atom. Nitrogen therefore has an oxidation state of –3 and hydrogen +1.

Rules to remember

1 The oxidation state of all elements uncombined is zero. Therefore, the oxidation state of oxygen in oxygen gas is zero.
2 The algebraic sum of the oxidation states of the elements in a compound is always zero.

For example, in SO_2

S	O.S.	+4
O	O.S.	–2
O	O.S.	–2
	Total	0

N.B. The electrons in the S=O bonds are assigned to oxygen because oxygen is more electronegative than sulphur.

3 The algebraic sum of the oxidation states of the elements in an ion is equal to the charge on the ion.

For example, in CO_3^{2-}

C	O.S.	+4
O	O.S.	–2
O	O.S.	–2
O	O.S.	–2
	Total	–2

4 The oxidation state of oxygen is – 2 (except in oxygen gas and peroxides).
5 The oxidation state of hydrogen is +1 (except when combined with Group 1 or 2 metals in hydrides).

Writing chemical names with oxidation states

In chemical names the oxidation state of a particular element may be shown in Roman numerals in brackets if there is the chance of any uncertainty in its oxidation state. If this is done, the sign is not given.

For example:
iron(III) chloride: iron is in oxidation state +3
iron(II) chloride: iron is in oxidation state +2
tetracarbonylnickel(0): nickel is in oxidation state 0

Recognising changes in oxidation state

If during a chemical reaction an atom or ion changes its oxidation state, then oxidation and reduction are taking place.

An increase in oxidation state is oxidation.

A decrease in oxidation state is reduction.

For example, the reaction of chlorine with iron(II) chloride to form iron(III) chloride, is
$$Cl_2(g) + 2FeCl_2(aq) \rightarrow 2FeCl_3(aq)$$

This could be written in two ionic half equations.
$$Cl_2(g) + 2e^- \rightarrow 2Cl^-(aq)$$
$$Fe^{2+}(aq) \rightarrow Fe^{3+}(aq) + e^-$$

The oxidation state of chlorine in Cl_2 is zero (because it is an element) and in Cl^- it is −1.

The chlorine is therefore reduced because the oxidation state is reduced from 0 to −1. Similarly, the oxidation state of iron(II) increases from +2 to +3. The iron is oxidised because there is an increase in oxidation state.

Variable oxidation states

Vanadium is a transition metal that has compounds with different oxidation states.

If zinc and dilute sulphuric acid are added to a solution of a vanadium(V) compound, which is yellow, a series of different solutions are formed.

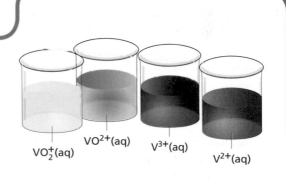

$VO_2^+(aq)$ $VO^{2+}(aq)$ $V^{3+}(aq)$ $V^{2+}(aq)$

The solutions are shown in the diagram and summarised in the table.

Oxidation state	Species	Colour
vanadium(V)	$VO_2^+(aq)$	yellow
vanadium(IV)	$VO^{2+}(aq)$	blue
vanadium(III)	$V^{3+}(aq)$	green
vanadium(II)	$V^{2+}(aq)$	mauve

Quick test

1 Work out the oxidation state of the element in **bold** type in each case.

a H_2 b **Cu**O   c **Cu**$_2$O   d H**Cl**   e Na**Cl**   f **S**O$_4^{2-}$
g I**O**$_3^-$ h K**Mn**O$_4$ I H$_2$**S**O$_4$ j Ca**C**O$_3$

2 What are the changes in oxidation state in this equation and which substances are oxidised and reduced?

$$Zn(s) + 2H^+(aq) \rightarrow Zn^{2+}(aq) + H_2(g)$$

1. (a) 0; (b) +2; (c) +1; (d) +1; (e) +1; (f) +6; (g) +5; (h) +7; (i) +6; (j) +4   2. Zn goes from 0 to +2 and so is oxidised. H+ goes from +1 to 0 and so is reduced.

53

The s-block elements

The s-block elements are the reactive metals of Groups 1 and 2 in the Periodic Table.
These are called the alkali metals (Group 1) and the alkaline earth metals (Group 2).
There are trends in the properties of the elements in these Groups.

Electron arrangements

The elements in Groups 1 and 2 are called s-block elements because their highest energy electrons are in an *s* subshell. An element in Group 1 has one electron more than a noble gas and an element in Group 2 has two more electrons.

EXAMINER'S TOP TIP
The electron arrangements in the table are shown in a simple form. [He] means the electron arrangement of a helium atom, i.e. $1s^2$. It is all right for you to use this abbreviation unless the question asks you to write electron arrangements in full.

Electron arrangement in elements in the s-block	
Group 1	Group 2
lithium Li [He]$2s^1$	beryllium Be [He]$2s^2$
sodium Na [Ne]$3s^1$	magnesium Mg [Ne]$3s^2$
potassium K [Ar]$4s^1$	calcium Ca [Ar]$4s^2$
rubidium Rb [Kr]$5s^1$	strontium Sr [Kr]$5s^2$
caesium Cs [Xe]$6s^1$	barium Ba [Xe]$6s^2$
francium Fr [Rn]$7s^1$	radium Ra [Rn]$7s^2$

The reactivity of these metals is explained because atoms of these elements easily lose 1 electron (Group 1 elements) or 2 electrons (Group 2 elements) to form ions which have electron arrangements similar to a noble gas.
For example:

sodium atom Na $\rightarrow$ sodium ion Na$^+$ + e$^-$
$1s^2 2s^2 2p^6 3s^1$ $\qquad$ $1s^2 2s^2 2p^6$

magnesium atom Mg $\rightarrow$ magnesium ion Mg^{2+} + 2e$^-$
$1s^2 2s^2 2p^6 3s^2$ $\qquad$ $1s^2 2s^2 2p^6$

Alkali metals and alkaline earth metals

Group 1 (alkali metals)
- are soft metals that can be cut with a knife
- have low melting and boiling points
- have low densities, for example, lithium, sodium and potassium float on water.

Group 2 (alkaline earth metals)
- are metals that have higher melting points than Group 1 metals
- have low densities, but not as low as Group 1.

Reaction of alkali metals and alkaline earth metals with water

The difference in reactivity of these elements can be seen in the reactions of these metals with cold water.
Lithium reacts slowly with cold water to produce the alkali lithium hydroxide and hydrogen.

$$2Li(s) + 2H_2O(l) \rightarrow 2LiOH(aq) + H_2(g)$$

Sodium reacting with water

Reactions continued

Sodium reacts quickly with cold water producing hydrogen. This does not usually catch alight. if it does, it burns with an orange yellow flame.

$$2Na(s) + 2H_2O(l) \rightarrow 2NaOH(aq) + H_2(g)$$

Potassium reacts rapidly with cold water and the hydrogen produced ignites spontaneously and burns with a pinkish-purple flame.

$$2K(s) + 2H_2O(l) \rightarrow 2KOH(aq) + H_2(g)$$

Rubidium and **caesium** react explosively with cold water.

In Group 2, **beryllium** does not react with water. **Magnesium** reacts very slowly with hot water but rapidly with steam.

$$Mg(s) + H_2O(g) \rightarrow MgO(s) + H_2(g)$$

(Magnesium hydroxide decomposes at high temperatures into magnesium oxide.)

Calcium, **strontium** and **barium** react with cold water, e.g.

$$Ca(s) + 2H_2O(l) \rightarrow Ca(OH)_2(aq) + H_2(g)$$

Reaction of alkali metals and alkaline earth metals with air

A variety of oxides are produced when these metals burn in oxygen. For Group 1 metals, the common oxide of formula M_2O formed by all, but the more reactive metals form additional oxides, e.g. potassium peroxide K_2O_2. Group 2 metals all form a typical oxide of formula MO but, in addition, strontium and barium form peroxides SrO_2 and BaO_2.

Reaction of alkali metals and alkaline earth metals with chlorine

The elements in Groups 1 and 2 react with chlorine on heating to produce the chlorides, e.g.

$$2Na(s) + Cl_2(g) \rightarrow 2NaCl(s)$$
$$Mg(s) + Cl_2(g) \rightarrow MgCl_2(s)$$

EXAMINER'S TOP TIP
You should have seen a pattern in properties of Group 1 metals at GCSE. Group 2 metals again show trends in properties.

The difference in reactivity
in Groups 1 and 2

In each group, the ionisation energies decrease and the chemical reactivities increase down the group. The ionisation energies are:

Li	520 kJ mol^{-1}	Be	2700 kJ mol^{-1}
Na	500 kJ mol^{-1}	Mg	2240 kJ mol^{-1}
K	420 kJ mol^{-1}	Ca	1690 kJ mol^{-1}
Rb	400 kJ mol^{-1}	Sr	1650 kJ mol^{-1}
Cs	380 kJ mol^{-1}	Ba	1500 kJ mol^{-1}

(N.B. First ionisation energies are given for Group 1 elements and the sum of first and second ionisation energies for Group 2.)

The decrease in ionisation energies down the group can be explained by the increasing atomic radius and increased shielding of filled electron shells down the group.

Quick test

1 Which is the least reactive alkaline earth metal?
2 The alkali metals and barium are usually stored in oil. Why is this?
3 Write a balanced equation for the reaction of lithium with oxygen.

1. Beryllium 2. To prevent the metals reacting with oxygen in the air and water. 3. $4Li(s) + O_2(g) \rightarrow 2Li_2O(s)$.

Compounds of s-block elements

Compounds of s-block elements are usually <u>white</u> or <u>colourless solids</u>.
There are trends in properties within Group 1 and Group 2.

Oxides

The oxides of Group 1 react rapidly when added to water to form strongly alkaline solutions, e.g.

$$Na_2O(s) + H_2O(l) \rightarrow 2NaOH(aq)$$

Lithium, however, does not form a strongly alkaline solution.

The tendency for Group 2 oxides to form alkaline solutions when added to water is less, e.g.

$$CaO(s) + H_2O(l) \rightarrow Ca(OH)_2(aq)$$

The basic strengths of the resulting solutions from both groups increases down each group. The ionic radii of the metal ions increase down each group. There is, therefore, less attraction between the M^+ or M^{2+} ion and the OH^- ion.

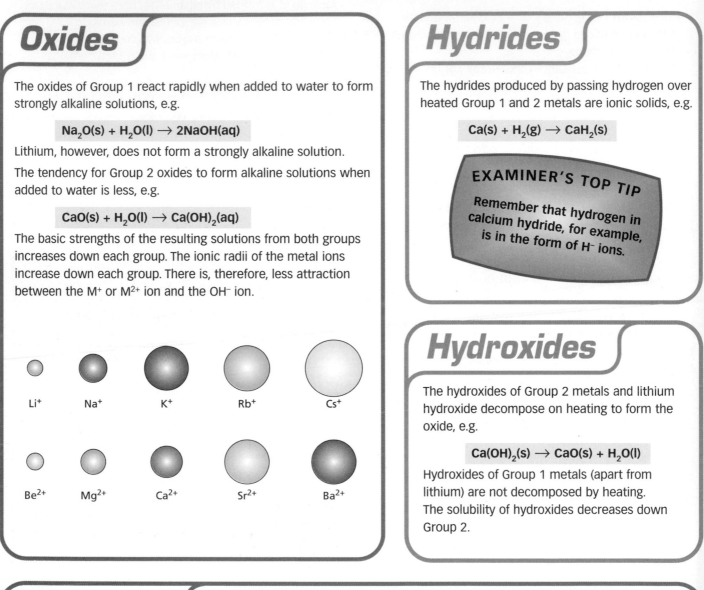

Li^+ Na^+ K^+ Rb^+ Cs^+

Be^{2+} Mg^{2+} Ca^{2+} Sr^{2+} Ba^{2+}

Hydrides

The hydrides produced by passing hydrogen over heated Group 1 and 2 metals are ionic solids, e.g.

$$Ca(s) + H_2(g) \rightarrow CaH_2(s)$$

EXAMINER'S TOP TIP

Remember that hydrogen in calcium hydride, for example, is in the form of H^- ions.

Hydroxides

The hydroxides of Group 2 metals and lithium hydroxide decompose on heating to form the oxide, e.g.

$$Ca(OH)_2(s) \rightarrow CaO(s) + H_2O(l)$$

Hydroxides of Group 1 metals (apart from lithium) are not decomposed by heating. The solubility of hydroxides decreases down Group 2.

Chlorides

The chlorides of Group 1 and 2 metals are generally ionic. They are
- **white, crystalline solids** with high melting points:
- **poor conductors** of electricity when solid but undergo electrolysis when molten or dissolved in water.

Beryllium chloride is different from the other chlorides of Group 2. Beryllium is more electronegative (or less electropositive) than the other elements of the Group. As a result it shows the greatest tendency to covalent bonding. It is:
- **a solid**
- **a poor conductor** of electricity when molten
- **soluble** in organic solvents
- **hydrolysed** when heated in aqueous solution.

The structure of beryllium chloride is:

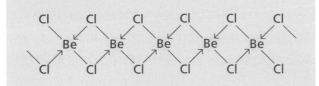

Carbonates

Carbonates of Group 1 metals dissolve in water to form an alkaline solution. The solubility of lithium carbonate is much less than that of the other Group 1 carbonates. Only lithium carbonate is decomposed by heating:

$$Li_2CO_3(s) \rightarrow Li_2O(s) + CO_2(g)$$

Group 2 metal carbonates are insoluble in water and are decomposed by heating into the oxide and carbon dioxide, e.g.

$$CaCO_3(s) \rightarrow CaO(s) + CO_2(g)$$

The decomposition takes place because of the polarising power of the 2+ ion.

The polarising power decreases down the Group.

Salts of Groups 1 and 2

Salts of Group 2 elements generally decompose more readily than salts of Group 1 elements, e.g. the decomposition of nitrate.

Salts of Group 1 metals are generally soluble in water. Some salts of lithium, however, are insoluble, e.g. Li_2CO_3 and LiF. This can be explained by the high lattice energies of these compounds caused by the small size of the lithium ion. The hydration enthalpy produced when the substance dissolves is much less than the lattice enthalpy.

Lithium fluoride:		
lattice enthalpy	=	1022 kJ mol^{-1}
hydration enthalpy	=	– 833 kJ mol^{-1}
lattice enthalpy	>	hydration enthalpy
Lithium fluoride is insoluble.		

Sodium fluoride:		
lattice enthalpy	=	902 kJ mol^{-1}
hydration enthalpy	=	– 912 kJ mol^{-1}
lattice enthalpy	<	hydration enthalpy
Sodium fluoride is soluble.		

The table shows the pattern in solubility of some sulphates of metals in Group 2.

Group 2 metal sulphate	Solubility in water in g/100 g of water
Magnesium	33
Calcium	0.21
Strontium	0.13
Barium	0.00024

Solubility decreases down the Group

Sulphates decrease in solubility down the Group. This is due to decreasing hydration energies down the Group.

Flame tests

Many salts of alkali metals and alkaline earth metals give characteristic colours in a flame test.

Metal present	Colour of flame
lithium	red
sodium	orange
potassium	pinkish purple (lilac)
calcium	brick red
barium	apple green

Quick test

1 *What evidence is there that beryllium chloride is not ionic?*

2 *Lithium and magnesium are similar in reacting with nitrogen to form nitrides.*
Write symbol equations for the reaction of lithium and magnesium with nitrogen.

3 *Calcium fluoride has a lattice enthalpy of 2602 kJ mol^{-1} and a hydration enthalpy of 2662 k mol^{-1}. Is it soluble in water? Explain your choice.*

1. Beryllium chloride is a poor conductor of electricity when molten. It is soluble in organic solvents and is hydrolysed when heated in aqueous solution   2. 4Li(s) + N$_2$(0) → 2Li$_2$N(s) 3Mg(s) + N$_2$(g) → Mg$_3$N$_2$(s)   3. Yes. Hydration enthalpy is greater than lattice enthalpy.

57

Halogens

There is pattern in the physical and chemical properties of the elements of Group 7 (called the halogens).

The halogens are a family of non-metals.   Halogens are good oxidising agents.

The halogen elements

The elements in Group 7 are shown in the table.

Element	Symbol	Electron arrangement	Appearance
fluorine	F	[He] $2s^2 2p^5$	pale greenish-yellow gas
chlorine	Cl	[Ne] $3s^2 3p^5$	greenish-yellow gas
bromine	Br	[Ar] $3d^{10} 4s^2 4p^5$	dark red liquid
iodine	I	[Kr] $4d^{10} 5s^2 5p^5$	grey-black solid

All of the halogen atoms are one electron short of a noble gas arrangement,

e.g.   **fluorine** $1s^2 2s^2 2p^5$ **neon** $1s^2 2s^2 2p^6$

Halogen atoms can combine by ionic and covalent bonding.

1 The halogen atoms can gain one electron to form negatively charged halide ions,

e.g. $Cl + e^- \rightarrow Cl^-$

2 A halogen atom can also form a covalent bond by overlap of the p-orbital containing one electron with a partially filled orbital on another atom.

Fluorine exhibits a maximum covalency of one. The other elements can exhibit covalencies of one, three, five or seven corresponding to the promotion of electrons into the available d orbitals.

For example, iodine can show higher covalencies.

One unpaired electron forms one covalent bond, e.g. ICl.
One electron may be promoted into a $5d$ orbital:

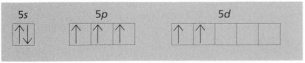

Three unpaired electrons form three covalent bonds, e.g. ICl_3.
Two electrons may be promoted into $5d$ orbitals:

Five unpaired electrons form five covalent bonds, e.g. ICl_5.
Three electrons may be promoted into $5d$ orbitals:

Seven unpaired electrons form seven covalent bonds, e.g. ICl_7.

Halogen molecules

All halogens are composed of diatomic molecules. Within the molecule, the two atoms are joined by a single covalent bond.

The enthalpies of atomisation of elements are:

fluorine	79.1 kJ mol^{-1}
chlorine	121.1 kJ mol^{-1}
bromine	112.0 kJ mol^{-1}
iodine	106.6 kJ mol^{-1}

From chlorine to iodine, the strength of the bond between the halogen atoms decreases. Fluorine has a much weaker bond than might be expected and this is because of the extra repulsion caused by the close proximity of the two fluorine nuclei and repulsion between non-bonding electrons on both atoms.

Reactions of halogens

Reaction with hydrogen

The reactions of hydrogen and halogens show the decrease in reactivity down Group 7.

Hydrogen and halogen mixture	Reaction with hydrogen
hydrogen and fluorine	explosive
hydrogen and chlorine	explosive in sunlight
hydrogen and bromine	reacts when mixture heated
hydrogen and iodine	reversible reaction when heated

Reactivity decreases down the Group.

EXAMINER'S TOP TIP
The reactivity of halogens decreases down the Group. This contrasts with the situation in Groups 1 and 2.

Reactions of halogen

Reactions with water

Fluorine reacts vigorously with cold water to form hydrogen fluoride and oxygen.

$$2F_2(g) + 2H_2O(l) \longrightarrow 4HF(g) + O_2(g)$$

Chlorine reacts less readily with water. A solution of chlorine in water (chlorine water) is acidic due to the formation of hydrochloric acid and chloric(I) acid (hypochlorous acid).

$$Cl_2(g) + H_2O(l) \rightleftharpoons HCl(aq) + HOCl(aq)$$

When this solution is exposed to strong sunlight the chloric(I) acid decomposes to produce oxygen.

$$2HOCl(aq) \longrightarrow 2HCl(aq) + O_2(g)$$

It is this available oxygen which accounts for the bleaching properties of chlorine.

Reactions with alkalis

Chlorine and bromine also react with aqueous potassium hydroxide solution, with the products again depending upon the conditions. With cold, dilute potassium hydroxide solution, potassium chloride and potassium chlorate(I) (potassium hypochlorite) are formed.

$$3ClO^-(aq) \longrightarrow 2Cl^-(aq) + ClO_3^-(aq)$$

The overall reaction is

$$Cl_2(g) + 2OH^-(aq) \longrightarrow Cl^-(aq) + ClO^-(aq) + H_2O(l)$$

With hot, concentrated potassium hydroxide, the chlorate(I) or bromate(I) disproportionates.

$$3ClO^-(aq) \longrightarrow 2Cl^-(aq) + ClO_3^-(aq)$$

chlorate (V)

Displacement reactions of halogens

The decrease in reactivity down Group 7 can be shown by displacement reactions.

A halogen element will replace a halogen in a salt providing it is more reactive than the halogen in the salt.

e.g. sodium iodide + chlorine $\longrightarrow$ iodine + sodium chloride

$$2NaI(aq) + Cl_2(g) \longrightarrow I_2(aq) + 2NaCl(aq)$$

$$2I^-(aq) + Cl_2(g) \longrightarrow I_2(aq) + 2Cl^-(aq)$$

EXAMINER'S TOP TIP
A disproportionation reaction involves a substance being both oxidised and reduced.

Quick test

1 For each of the following state whether there is an increase or a decrease down Group 7.

   a Atomic radius
   b Ionic radius
   c Boiling point
   d Electronegativity
   e Reactivity

2 Bromine is added to potassium chloride solution and potassium iodide solution.

   Is there a reaction in each case? Explain your answers. Write symbol and ionic equations for any reactions that take place.

1. (a) increase; (b) increase; (c) increase; (d) decrease; (e) decrease   2. Bromine and potassium chloride solution – no reaction. Bromine is less reactive than chlorine. Bromine and potassium iodide solution – reaction. Bromine is more reactive than iodine.

$$2NaI(aq) + Br_2(g) \longrightarrow I_2(aq) + 2NaBr(aq)$$

$$2I^-(aq) + Br_2(g) \longrightarrow I_2(aq) + 2Br^-(aq)$$

Halogen compounds

The halogen elements in Group 7 form compounds with nearly all elements.
Even noble gases that do not usually form compounds combine with halogens.
These compounds can contain ionic or covalent bonding.

Halides

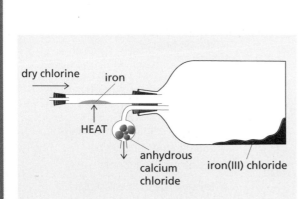

dry chlorine
iron
HEAT
anhydrous
calcium
chloride
iron(III) chloride

The halides formed by electropositive metals (e.g. sodium) are ionic. Halides of less electropositive metals and non metals are covalent. Where an element forms two chlorides, the one with the element in the higher oxidation state is more covalent, e.g. $SnCl_4$ (the oxidation state of tin is +4) is more covalent than $SnCl_2$ (oxidation state +2).

Preparing halides

Anhydrous halides are prepared by passing dry halogen vapour over the heated element.

e.g. $$2Fe(s) + 3Cl_2(g) \rightarrow 2FeCl_3(s)$$
iron(III) chloride

N.B. The reaction of iron with dry hydrogen chloride produces iron(II) chloride.

$$Fe(s) + 2HCl(g) \rightarrow FeCl_2(s) + H_2(g)$$

Reactions of halides with concentrated sulphuric acid

<u>**Concentrated sulphuric**</u> acid acts as an oxidising agent. The increasing reducing power down the group leads to different products.

Fluoride and chloride

Hydrogen fluoride or hydrogen chloride are produced.

$$NaCl(s) + H_2SO_4(l) \rightarrow NaHSO_4(s) + HCl(g)$$

Hydrogen fluoride or hydrogen chloride are not strong enough reducing agents for a redox reaction to take place.

Bromide

Hydrogen bromide is formed initially and then hydrogen bromide reduces the sulphuric acid to sulphur dioxide. Hydrogen bromide is oxidised to brown fumes of bromine.

$$NaBr(s) + H_2SO_4(l) \rightarrow NaHSO_4(s) + HBr(g)$$
$$2HBr(g) + H_2SO_4(l) \rightarrow SO_2(g) + Br_2(g) + 2H_2O(l)$$

Here, sulphur is reduced from oxidation state +6 to +4 and bromine is oxidised from −1 to 0.

Iodide

Hydrogen iodide is formed initially and then hydrogen iodide reduces the sulphuric acid first to sulphur dioxide and then hydrogen sulphide.

Hydrogen halides

Hydrogen halides can be produced by direct combination.

e.g. $$H_2(g) + Cl_2(g) \rightarrow 2HCl(g)$$

They are colourless gases at room temperature. They are very soluble in water forming strongly acidic solutions. For example, hydrogen chloride dissolves in water to form hydrochloric acid:

$$HCl(g) + aq \rightarrow H^+(aq) + Cl^-(aq)$$

Hydrogen fluoride, however, is only a weak acid and is only partially dissociated in water. Down the group the acidity increases because the H–X bond enthalpy decreases down the group.

Hydrogen halides are reducing agents with the reducing ability increasing down the group.

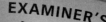

Hydrolysis of halides

Most ionic halides are not hydrolysed (split up) by water and can be produced by reacting a metal, metal oxide, metal hydroxide or metal carbonate with a hydrogen halide, followed by evaporation and crystallisation, e.g.

$$NaOH(aq) + HCl(aq) \rightarrow NaCl(aq) + H_2O(l)$$

EXAMINER'S TOP TIP
Metal halides containing water of crystallisation are often hydrolysed on heating. It is not possible to produce an anhydrous metal halide by heating a hydrated metal halide.

Tests for halides

Chlorides, bromides and iodides can be tested for using dilute nitric acid and silver nitrate solution.

e.g. $$Ag^+(aq) + Cl^-(aq) \rightarrow AgCl(s)$$

The colour of the precipitate can be used to distinguish which halogen is present.

Silver halide	Formula	Colour of precipitate
silver chloride	AgCl	white
silver bromide	AgBr	cream
silver iodide	AgI	yellow

The silver halide precipitates can be distinguished by their solubilities in ammonia solution.

$$AgCl(s) + 2NH_3(aq) \rightarrow [Ag(NH_3)_2]^+(aq) + Cl^-(aq)$$

Silver chloride is soluble, silver bromide is partially soluble and silver iodide is virtually insoluble in excess ammonia solution.

Fluorides cannot be tested for using silver nitrate solution as silver fluoride is soluble in water.

This is because the fluoride ion has a particularly high hydration enthalpy due to its small size.

Quick test

1 Write equations for the reduction of sulphuric acid with hydrogen iodide first to sulphur dioxide and then to hydrogen sulphide. What changes in oxidation state are there in each case?

2 How could you distinguish the following pairs of compounds?
  a Sodium chloride and sodium bromide.
  b Sodium fluoride and sodium bromide.

3 When testing for chloride, bromide or iodide, the solution should be acidified with dilute nitric acid. Suggest why this is so.

1. $2HI(g) + H_2SO_4(l) \rightarrow SO_2(g) + I_2(g) + 2H_2O(l)$ Oxidation state of S from +6 to +4, i.e. a reduction; oxidation state of I from −1 to 0, i.e. an oxidation. $6HI(g) + H_2SO_4(l) \rightarrow H_2S(g) + 3I_2(s) + 2H_2O(l)$ Oxidation state of S from +6 to −2, i.e. a reduction; oxidation state of I from −1 to 0, i.e. an oxidation. 2. (a) Add dilute nitric acid and silver nitrate solution. A white precipitate of silver chloride soluble in ammonia solution confirms chloride. A cream precipitate partially soluble in ammonia solution confirms bromide; (b) Add dilute nitric acid and silver nitrate solution. No white precipitate with fluoride. A cream precipitate partially soluble in ammonia solution confirms bromide. Fluoride forms colourless hydrogen fluoride. Or: add concentrated sulphuric acid. Fluoride forms colourless hydrogen fluoride. Bromide forms hydrogen bromide which is oxidised to red bromine fumes. 3. Without acidification with dilute nitric acid, carbonate could be present and a white precipitate of silver carbonate be formed. This could be confused with silver chloride.

Industrial processes

The chemical industry uses chemical principles to achieve the best conditions for processes to produce new chemicals.

It is important that industry produces chemicals as cheaply as possible.

Electrolysis of brine (sodium chloride solution)

Brine is obtained by pumping water underground in areas such as Cheshire where there are underground salt deposits. The salt dissolves to form brine.

Electrolysis of sodium chloride solution produces **sodium hydroxide, chlorine** and **hydrogen**.

The diagram shows a membrane cell.

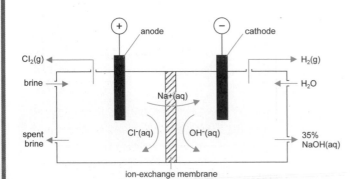

Left-hand compartment
Purified brine enters the cell.
Chlorine is produced at the anode.

$$2Cl^-(aq) \rightarrow Cl_2(g) + 2e^-$$

Sodium ions pass through the ion-exchange membrane.

Right-hand compartment
Water enters the cell.
Hydrogen is produced at the cathode.

$$2H_2O(l) + 2e^- \rightarrow 2OH^-(aq) + H_2(g)$$

Sodium hydroxide solution leaves the cell.

The ion-exchange membrane keeps the chlorine and sodium hydroxide apart to prevent reaction.

EXAMINER'S TOP TIP
Electrolysis in industry is expensive because it uses large amounts of electricity. In this process three products are formed that have plenty of uses and are easy to sell.

Reaction of chlorine with sodium hydroxide

Chlorine and sodium hydroxide react together to form sodium chlorate(I) which is used as household bleach (see page 59).

The strength of a household bleach can be found by titration.

Strength of a bleach
Excess potassium iodide solution is added to a sample of bleach. Chlorine in the bleach is a stronger oxidising agent than iodine. Iodine is released.

$$Cl_2(aq) + 2I^-(aq) \rightarrow 2Cl^-(aq) + I_2(aq)$$

Sodium thiosulphate solution is added from a burette until the solution is pale straw colour. Starch indicator is added and the solution goes dark blue.

Sodium thiosulphate solution is then added until the solution is just colourless

$$S_2O_3^{2-}(aq) + 2I^-(aq) \rightarrow S_4O_6^{2-}(aq) + I_2(aq)$$

Manufacture of ammonia

The growth in population since the start of the 20th century made it important to devise a process that would 'fix' nitrogen in the atmosphere to enable nitrogen fertilisers to be made in large quantities.

The graph shows how the **growth in population** has been matched by **ammonia production**.

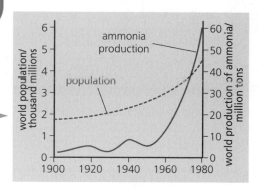

The process was developed by the German chemist Fritz Haber. It is called the **Haber Process**.

Getting the reactants for the process

Hydrogen is produced for this process by passing methane or naphtha mixed with steam over a heated nickel catalyst at pressures up to 30 atm. The products are **carbon monoxide** and **hydrogen**. The gases produced are mixed with steam and passed over a **heated catalyst** to produce **carbon dioxide** and **hydrogen**. The carbon dioxide is **dissolved in water** under pressure. Nitrogen for the industrial process is produced by **fractional distillation** of liquid air.

The diagram summarises the Haber process.

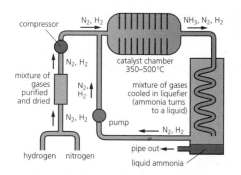

The equation for the Haber process is:

$$N_2(g) + 3H_2(g) \rightleftharpoons 2NH_3(g) \qquad \Delta H = -92 \text{ kJ mol}^{-1}$$

Getting the best conditions

In the Haber process, 2 moles of ammonia are formed from 4 moles of gas (mixture of 1 mole of nitrogen and 3 moles of hydrogen). According to Le Chatalier's principle:

1 If the **pressure is increased**, the **equilibrium** will move to the right to give a **smaller volume** and hence produce **more ammonia**.

2 **More ammonia** is produced if the **temperature is decreased**. The equilibrium should move in the **exothermic** direction to compensate for the decrease in temperature.

Theoretically, high pressures and low temperatures should give the best yield of ammonia, but:

- Unreacted gases can be recycled so the equilibrium does not have to be reached.
- High pressure equipment becomes increasingly expensive as pressure increases.
- At low temperatures reactions are very slow. Even using a catalyst to speed up the reaction may not make the process fast enough to be economical.
- At low temperatures catalyst life and activity are prolonged.

What are the best conditions?

The reaction conditions are a compromise and the usual conditions are:

(i) a pressure of 200 atmospheres;

(ii) a temperature around 380°–450°C;

(iii) a catalyst of finely divided iron containing promoters to stop the catalyst being poisoned.

The percentage conversion under these conditions is about 15% and the equilibrium position is never reached in the converter.

EXAMINER'S TOP TIP
Students frequently give the atmosphere as a source of hydrogen. There is no hydrogen normally in the atmosphere.

Quick test

1 *Write equations for the reactions used to produce hydrogen in the Haber process.*

2 *Suggest what factors might be important in choosing a site for a new ammonia factory.*

3 *Ammonia is used to make nitric acid. Ammonia and air (containing oxygen) react to form nitrogen monoxide and steam. On cooling nitrogen monoxide combines with more oxygen to form nitrogen dioxide. Dissolving nitrogen dioxide in water, in the presence of air, forms nitric acid. Write symbol equations for these reactions.*

4 *Ammonia is converted into ammonium nitrate and ammonium sulphate.*

Write equations for these reactions.

1. $CH_4(g) + H_2O(g) \longrightarrow CO(g) + 3H_2(g)$
$CO(g) + H_2O(g) \longrightarrow CO_2(g) + H_2(g)$ 2. sources – coal, gas or oil, large quantities of water, suitable transport system, markets close by.   3. $4NH_3(g) + 5O_2(g) \longrightarrow 4NO(g) + 6H_2O(g)$ $2NO(g) + O_2(g) \longrightarrow 2NO_2(g)$
$4NO_2(g) + 2H_2O(l) + O_2(g) \longrightarrow 4HNO_3(l)$ 4. $HNO_3(aq) + NH_3(aq) \longrightarrow NH_4NO_3(aq)$ $H_2SO_4(aq) + 2NH_3(aq) \longrightarrow (NH_4)_2SO_4(aq)$

63

Extraction of metals – 1

Metals are extracted from their <u>ores</u> by a variety of <u>oxidation</u> and <u>reduction</u> methods. The particular method used to extract a metal depends upon the <u>reactivity of the metal</u>. The most reactive metals are extracted by <u>electrolysis</u>.

Extraction of aluminium

Aluminium is found in the Earth as the ore bauxite, $Al_2O_3.2H_2O$. Bauxite contains a lot of impurities such as iron(III) oxide. It is purified before it is used for extracting aluminium.

Purifying bauxite

The ore is heated with sodium hydroxide solution under pressure. The aluminium oxide reacts to form the aluminate ion which remains in solution. The solid impurities are removed by filtration.

$$Al_2O_3(s) + 2OH^-(aq) + 3H_2O(l) \longrightarrow 2[Al(OH)_4]^-(aq)$$

Some freshly prepared aluminium hydroxide is added to 'seed' the solution of sodium aluminate and precipitate aluminium hydroxide.

$$[Al(OH)_4]^-(aq) \longrightarrow Al(OH)_3(s) + OH^-(aq)$$

The aluminium hydroxide is heated to produce alumina (aluminium oxide).

$$2Al(OH)_3(s) \longrightarrow Al_2O_3(s) + 3H_2O(l)$$

EXAMINER'S TOP TIP
You would not have studied the purification of the ore for GCSE. The purification of the ore relies on the amphoteric properties of aluminium oxide.

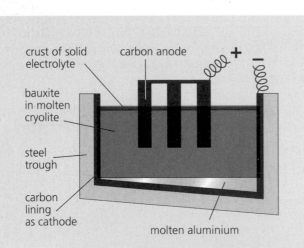

crust of solid electrolyte
carbon anode
bauxite in molten cryolite
steel trough
carbon lining as cathode
molten aluminium

Extraction of aluminium

Electrolysis of aluminium oxide

The extraction of aluminium is carried out by electrolysis of molten aluminium oxide dissolved in molten cryolite (sodium aluminium fluoride).

Electrolysis takes place in carbon-lined steel tanks called pots. The carbon lining acts as a cathode and carbon anodes are used.

The products are aluminium (produced at the cathode) and oxygen (produced at the anode).

Cathode: $Al^{3+} + 3e^- \longrightarrow Al$

aluminium ions + electrons $\longrightarrow$ aluminium atoms

Anode: $2O^{2-} \longrightarrow O_2 + 4e^-$

oxide ions $\longrightarrow$ oxygen molecules + electrons

The aluminium collects at the bottom of the pot and can be removed. The carbon anodes burn in the oxygen produced and have to be replaced from time to time. The exhaust gases from the pots are bubbled through water to remove soluble gases that contain fluorides.

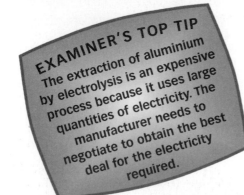

EXAMINER'S TOP TIP
The extraction of aluminium by electrolysis is an expensive process because it uses large quantities of electricity. The manufacturer needs to negotiate to obtain the best deal for the electricity required.

Oxidation and reduction

The diagram summarises the relationship between extraction of a metal (reduction) and corrosion of a metal (oxidation).

metal extraction
(reduction)

metal oxide $\longrightarrow$ metal

corrosion
(oxidation)

Extraction of titanium

Titanium is an important metal. It is a very strong metal which is less dense than steel, does not corrode and has a high melting point. It is, however, expensive to extract.

It cannot be extracted by reduction with carbon because this leaves traces of carbon in the metal as carbides. This makes the metal brittle.

Titanium is found in the Earth in the ores rutile, TiO_2, and ilmenite, $FeTiO_3$.

Purifying the ore produces pure titanium(IV) oxide. This is heated with carbon in a stream of chlorine.

Titanium(IV) chloride is formed.

$$TiO_2(s) + 2Cl_2(g) \rightarrow TiCl_4(l) + CO_2(g)$$

Titanium(IV) chloride can be purified by fractional distillation.

Titanium(IV) chloride is then heated with sodium at 500 °C in an atmosphere of argon. Exactly the right amount of sodium is added to react with the chlorine.

$$TiCl_4(g) + 4Na(s) \rightarrow Ti(s) + 4NaCl(s)$$

After cooling, dilute hydrochloric acid is added to remove the sodium chloride and leave titanium metal, which is then washed and dried.

Blades of aeroengine fan are made of titanium to withstand high temperatures.

Quick test

1 The extraction of aluminium is a continuous process but the extraction of titanium is a batch process.
   Why is a batch process more expensive to run than a continuous process?

2 Suggest why the extraction of titanium is an expensive process.

3 How do the manufacturers ensure that the titanium they produce is pure?

4 Magnesium can be used instead of sodium in the titanium extraction process.
   Write symbol equations for the reaction.

5 Sodium is extracted by the electrolysis of molten sodium chloride.
   Write equations for the reactions at the two electrodes.

1. In a batch process products are not made all the time so less is produced. 2. Sodium is an expensive metal as it uses electrolysis in its extraction. It is used in the extraction of titanium. There are also a number of stages in the process. It is also a batch process. 3. Titanium(IV) chloride is purified by fractional distillation and exactly the correct amount of sodium is used to remove all the chloride. 4. $TiCl_4(l) + 2Mg(s) \rightarrow Ti(s) + 2MgCl_2(s)$ 5. Cathode: $Na^+ + e^- \rightarrow Na$ Anode: $2Cl^- \rightarrow Cl_2 + 2e^-$

65

Extraction of metals – 2

- More <u>iron</u> is extracted from the <u>Earth</u> than any other metal.
- Most of the iron is converted into <u>steel</u>.

Extraction of iron

Iron exists in the Earth's crust in ores such as haematite (Fe_2O_3), magnetite (Fe_3O_4) and iron pyrites (FeS_2). Iron is obtained by reducing the ores in a **blast furnace**.

The blast furnace is loaded with a charge of **iron ore**, **coke** and **limestone** through the top of the furnace. The furnace is heated by blasts of hot air into the base. Various reactions take place in the furnace.

- The burning of the coke in the air produces temperatures in excess of 1500 °C.

$$C(s) + O_2(g) \rightarrow CO_2(g) \qquad \Delta H = -394 \text{ kJ mol}^{-1}$$

- The reduction of carbon dioxide to carbon monoxide.

$$CO_2(g) + C(s) \rightarrow 2CO(g) \qquad \Delta H = +173 \text{ kJ mol}^{-1}$$

- The reduction of iron ore takes place in the furnace. At the top of the furnace

$$Fe_2O_3(s) + 3CO(g) \rightarrow 2Fe(l) + 3CO_2(g) \qquad \Delta H = -27 \text{ kJ mol}^{-1}$$

$$Fe_2O_3(s) + CO(g) \rightarrow 2FeO(s) + CO_2(g)$$

lower in the furnace where the temperature is higher

$$FeO(s) + C(s) \rightarrow Fe(l) + CO(g)$$

- The limestone is added to the furnace to remove impurities of silicon(IV) oxide in the ore. The calcium carbonate decomposes to form calcium oxide.

$$CaCO_3(s) \rightarrow CaO(s) + CO_2(g)$$

- The calcium oxide reacts with the silicon(IV) oxide to form calcium silicate (slag).

$$CaO(s) + SiO_2(s) \rightarrow CaSiO_3(l)$$

The slag floats on top of the molten iron and both can be tapped off separately.

Slag, the waste material in the process, is used as a phosphorus fertiliser or is used for chippings for road surfacing. Blowing air through molten slag produces a fibrous material called rock wool which is used as an insulating material in roofs and walls.

The iron produced is called **pig iron** and contains about 4% carbon plus other impurities such as phosphorus (0.1%), silicon (0.2–0.3%) and sulphur (0.05%). Pig iron is brittle and not suitable for a wide range of uses.

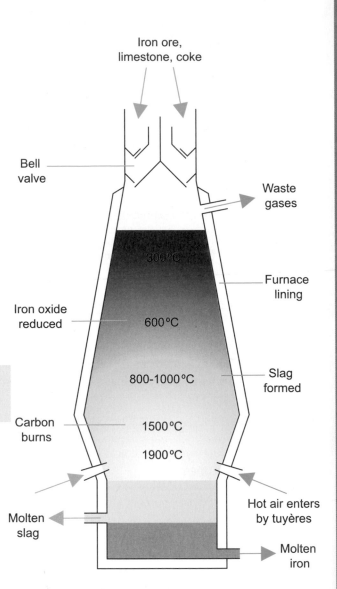

Steel making

Most of the pig iron is converted into steel. Steel is an **alloy** containing usually **0.5–1.5% carbon**.

The **molten iron** from the blast furnace is loaded into the steel making furnace with about half the quantity of scrap iron.

A water-cooled lance is then lowered into the upright furnace and pure oxygen, under high pressure, is blown onto the surface of the iron.

The **oxides of carbon**, **phosphorus** and **sulphur** produced escape from the furnace as gases.

Silicon is oxidised to **silicon(IV) oxide**. **Limestone** is added to react with this and form slag which can be tapped off.

Calculated quantities of carbon and other elements required in the steel are added to the furnace to give the type of steel required.

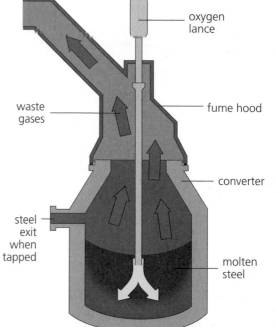

Types of steel

Type of steel	Composition	Uses
mild steel	about 0.2% carbon	car bodies
high carbon steel	about 2% carbon	rail tracks and tram lines
stainless steel	small amount of carbon and chromium and nickel	cutlery and saucepans
tool steel	small amount of carbon and tungsten	drill bits and cutting tools

Steel is an alloy. A metal consists of a closely packed structure of metal ions held together by a sea of electrons. The carbon atoms go into the gaps between ions. When these holes are filled, the structure is distorted and the steel is brittle.

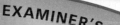

EXAMINER'S TOP TIP

This is an example of where a good understanding of a model helps to explain properties.

Quick test

1 Identify exothermic and endothermic reactions in the blast furnace.

2 Give two reasons for adding limestone to the furnace.

3 Steel from the blast furnace contains carbon, sulphur, phosphorus and silicon.
Write equations for the reactions removing these impurities in the furnace.

4 In the steel making process all of the impurities are removed and then measured amounts of carbon and other materials added. Why is this done?

5 The blast furnace is a continuous process and the steel making process is a batch process. Why is it an advantage to make steel by a batch process?

1. **exothermic:** $C(s) + O_2(g) \rightarrow CO_2(g)$; $Fe_2O_3(s) + 3CO(g) \rightarrow 2Fe(l) + 3CO_2(g)$ **endothermic:** $CO_2(g) + C(s) \rightarrow 2CO(g)$
2. to remove silicon(IV) oxide as slag; to decompose to form carbon dioxide which will make more carbon monoxide (active reducing agent). 3. $C(s) + O_2(g) \rightarrow CO_2(g)$ $S(s) + O_2(g) \rightarrow SO_2(g)$ $4P(s) + 5O_2(g) \rightarrow P_4O_{10}(g)$ $CaO(s) + SiO_2(s) \rightarrow CaSiO_3(l)$ 4. It is the way of knowing exactly how much of each element is present. 5. Different steels, with different compositions, have to be made. Each can be made in a separate batch.

Exam-style questions
Use the questions to test your progress. Check your answers on pages 92–93.

Inorganic chemistry

Use data from the Data section (pages 90–91) to help you answer these questions.

1 This question concerns the chlorides of the elements of Period 3 of the Periodic Table.
  a Complete the table. [8]

Na	Mg	Al	Si	P	S	Cl	Ar
	MgCl$_2$					Cl$_2$	—

  b Select one of the chlorides above and outline a laboratory method for its preparation. [3]

....

....

....

  c Write equations for the changes which occur when sodium chloride and phosphorus trichloride are added to water. [2]

....

....

2 This question is about the elements in Group 7 of the Periodic Table.
  a Complete the following table. [3]

Halogen	Physical state at room temperature	Colour
fluorine	gas	yellowish with greenish tinge
chlorine		
bromine		
iodine		

  b When potassium chloride is treated with concentrated sulphuric acid, a steamy gas, Y, is evolved.
  (i) Write an equation for this reaction. [2]

....

  (ii) When Y is dissolved in water, in which it is readily soluble, a strongly acidic solution, Z, is formed. Describe the reaction that occurs between Y and water in terms of the Brønsted–Lowry theory. [2]

....

....

  (iii) Silver nitrate solution is added to Z. A white precipitate forms. If silver in concentrated aqueous ammonia is carefully added to Z, no precipitate forms.
  Explain your observations. [2]

....

....

  c Aqueous chlorate(I) ions, ClO⁻, decompose on warming in a disproportionation reaction. Write the equation for this reaction and, by considering the oxidation state changes involved, explain the term 'disproportionation'. [3]

....

....

3 Iron occurs in the Earth's crust as several different compounds, but only a few are used as ores from which iron can be extracted. Two such compounds are the minerals haematite and magnetite, both of which are oxides of iron.

    a 2.32g of a pure sample of magnetite contained 1.68 g of iron. Calculate the formula of this oxide of iron. [2]

    b Iron is extracted from iron ore Fe_2O_3 in the blast furnace. Write an equation, including state symbols, for the reaction between this compound and carbon monoxide. [3]

....

....

    c Explain why calcium carbonate is added to the furnace. [2]

....

....

    d Sand for glassmaking must be as free as possible from iron compounds as impurities. Sand is made of silicon(IV) oxide, SiO_2, and iron is usually present as a thin coating of iron(III) oxide, Fe_2O_3, on each grain of sand. Use your knowledge of chemistry to suggest a method of removing the iron oxide coating from the sand grains. [2]

....

....

4 Use the graph of first ionisation energies of the elements from Na to Ar to answer the following questions.

    a Mark on the graph the approximate value of the ionisation energy of potassium, K. [1]

    b Explain the variation of first ionisation energy in descending a group. [1]

....

....

....

    c Explain how the graph provided evidence from electron arrangements in *s* and *p* levels. [2]

....

....

5 State what is oxidised and what is reduced in the following reactions.

    a $Zn(s) + 2HCl(aq) \rightarrow ZnCl_2(aq) + H_2(g)$ [2]

....

    b $2CuCl(aq) \rightarrow Cu(s) + CuCl_2(s)$ [2]

....

    c $PbO_2(s) + 4HCl(aq) \rightarrow PbCl_2(s) + 2H_2O(l) + Cl_2(g)$ [2]

....

    d $Cr_2O_3(s) + 2Al(s) \rightarrow Al_2O_3(s) + 2Cr(s)$ [2]

....

    e $3Cu(s) + 8HNO_3(aq) \rightarrow 3Cu(NO_3)_2 + 2NO(g) + 4H_2O(l)$ [2]

....

Total: /48

Organic chemistry

Organic chemistry is the study of carbon compounds excluding simple compounds such as carbon dioxide and carbonates.

There are more compounds of carbon than all other elements put together.

Organic compounds are made up from only a few elements.

Most organic compounds are made up from

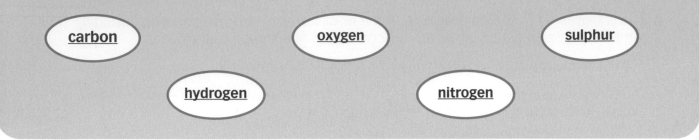

(carbon)   (oxygen)   (sulphur)

(hydrogen)   (nitrogen)

Why so many organic compounds?

Carbon atoms are able to form strong covalent bonds with other atoms. It is also possible to form rings. Silicon, an element similar to carbon, can only form compounds with up to nine silicon atoms in a chain.

Homologous series

Organic compounds can be put into families. A family of compounds is called an **homologous series**.

For example, four compounds in the same homologous series are:

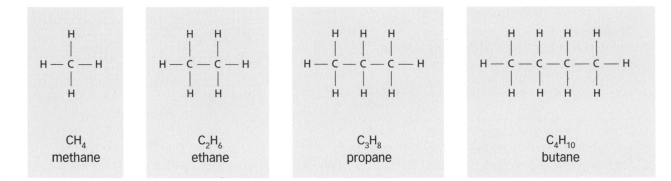

These four compounds are called **alkanes**. They all have the same **general formula** C_nH_{2n+2}.

These compounds have similar chemical properties and show a gradual change in physical properties.

alkanes	boiling point (°C)
methane	−162
ethane	−89
propane	−42
butane	−0.5

Each member of the homologous series differs by a unit of CH_2.

Stages in identifying an organic compound

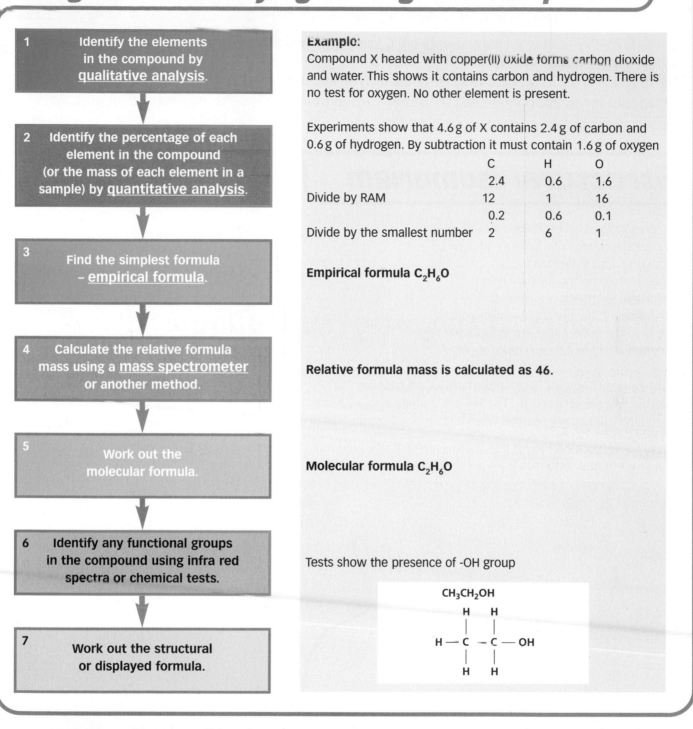

1	Identify the elements in the compound by <u>qualitative analysis</u>.

Example:
Compound X heated with copper(II) oxide forms carbon dioxide and water. This shows it contains carbon and hydrogen. There is no test for oxygen. No other element is present.

2	Identify the percentage of each element in the compound (or the mass of each element in a sample) by <u>quantitative analysis</u>.

Experiments show that 4.6 g of X contains 2.4 g of carbon and 0.6 g of hydrogen. By subtraction it must contain 1.6 g of oxygen

	C	H	O
	2.4	0.6	1.6
Divide by RAM	12	1	16
	0.2	0.6	0.1
Divide by the smallest number	2	6	1

Empirical formula C_2H_6O

3	Find the simplest formula – <u>empirical formula</u>.

4	Calculate the relative formula mass using a <u>mass spectrometer</u> or another method.

Relative formula mass is calculated as 46.

5	Work out the molecular formula.

Molecular formula C_2H_6O

6	Identify any functional groups in the compound using infra red spectra or chemical tests.

Tests show the presence of -OH group

CH_3CH_2OH

7	Work out the structural or displayed formula.

Quick test

1 Which of the following belongs to the homologous series of alkanes.

C_7H_{14} C_9H_{20} $C_{22}H_{44}$ $C_{30}H_{58}$

2 The following compounds are in the same homologous series.

C_2H_4O C_3H_6O C_4H_8O

  a Write down the general formula of the compounds in this list.

  b Predict the molecular formula of the compound in this homologous series containing 10 carbon atoms.

3 Predict the boiling point of the alkane containing 5 carbon atoms.

Isomerism

There are often different compounds with the same molecular formula but different structural formulae. These are called <u>structural isomers</u>.

Sometimes, atoms can be arranged differently in space leading to different <u>stereoisomers</u>.

Restricted rotation about a double carbon–carbon bond can lead to different <u>geometric isomers</u>.

Structural isomerism

It is possible for different compounds to have the same molecular formula but have different structural formulae.

These different compounds are called **structural isomers**.

One example is butane and 2-methylpropane. Both have the same molecular formula of C_4H_{10}. The structural formulae are shown opposite.

As the carbon chain gets longer more isomers are possible.

There are two isomers of C_4H_{10} but 366 319 isomers of $C_{20}H_{42}$.

Again more isomers are possible when functional groups are also present.

The table includes other examples.

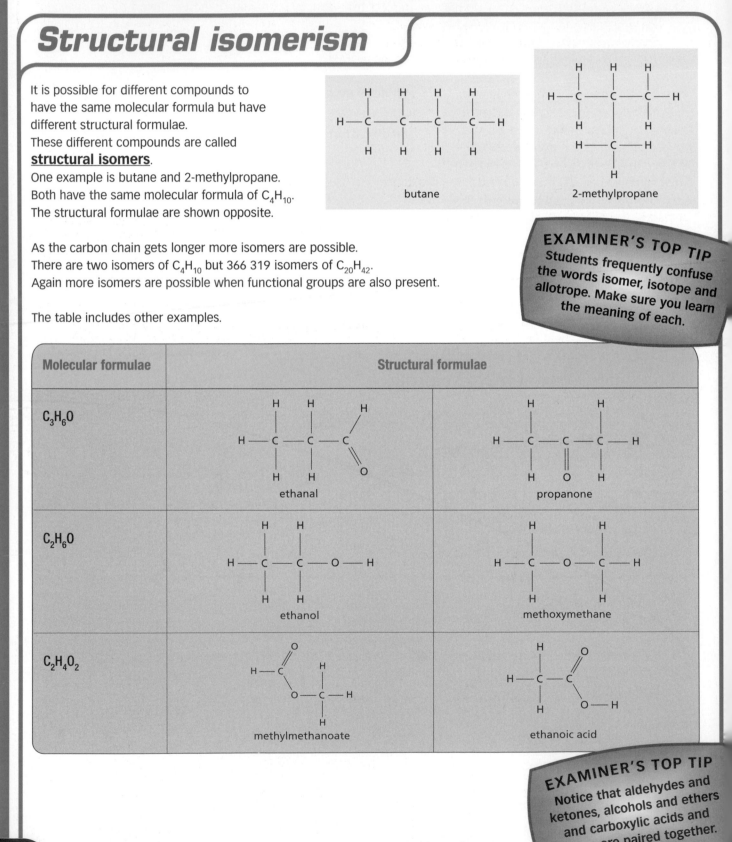

butane

2-methylpropane

EXAMINER'S TOP TIP
Students frequently confuse the words isomer, isotope and allotrope. Make sure you learn the meaning of each.

Molecular formulae	Structural formulae	
C_3H_6O	ethanal	propanone
C_2H_6O	ethanol	methoxymethane
$C_2H_4O_2$	methylmethanoate	ethanoic acid

EXAMINER'S TOP TIP
Notice that aldehydes and ketones, alcohols and ethers and carboxylic acids and esters are paired together.

Geometric isomers

There is normally free rotation about a carbon–carbon single bond but there is no free rotation about a carbon–carbon double or triple bond. The butenedioic acids give a common example of geometric isomerism. There are two possible isomers.

Lack of rotation in the carbon–carbon double bond explains the existence of the two isomers. The isomer with the two functional groups on the same side of the molecule is called the *cis* form. When the functional groups are on opposite sides of the molecule it is called the *trans* form.

The two isomers have different physical and chemical properties. The *cis* form melts at a lower temperature and loses a molecule of water to form the anhydride, because the two carboxyl groups are close together. The *trans* form loses a molecule of water only at a much higher temperature and the anhydride formed is identical to the anhydride formed from the *cis* form. At this higher temperature rotation has presumably taken place.

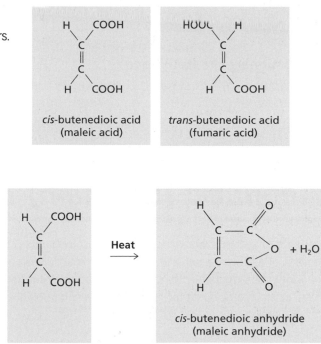

cis-butenedioic acid
(maleic acid)

trans-butenedioic acid
(fumaric acid)

Heat

cis-butenedioic anhydride
(maleic anhydride)

Geometric isomers in inorganic compounds

It is possible to get examples of geometric isomerism in inorganic chemistry.

For example, there are two geometric isomers of N_2F_2 due to no free rotation around a N=N bond.

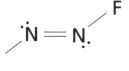

Quick test

1 Draw out and name the three structural isomers of C_5H_{12}.

2 Draw out possible isomers of C_3H_8O.

3 What isomers are there of the formula $C_2H_2Cl_2$?

3. two structural isomers:

1,1-dichloroethene
$CCl_2=CH_2$

1,2-dichloroethene
$CHCl=CHCl$

two geometric isomers:

cis-1,2-dichloroethene

trans-1,2-dichloroethene

2.

propan-1-ol
(alcohol)

pentane

propan-2-ol
(alcohol)

2-methylbutane

methoxyethane
(ether)

2,2-dimethylpropane.

1.

Naming organic compounds

- Systematic naming of organic compounds is needed because there are so many possible compounds.
- The system identifies the longest chain within the compound and uses this as the basis of naming the compound.
- Many common organic compounds still retain a common name which is not related to the chemical composition, e.g. propanone is still frequently called acetone.
- Since 1948, attempts have been made to systematise the names of all organic compounds.

Aliphatic (or chain) compounds

The basis of systematic naming of aliphatic compounds is that every name consists of a root, one suffix and as many prefixes as necessary.
The root is determined by the number of carbon atoms in the longest continuous chain. In the examples on the right, the longest carbon chain in A is 3 and in B is 4.

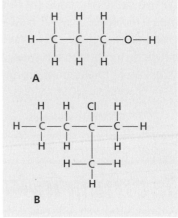

The table lists the alkanes from which roots are obtained.

Number roots for aliphatic nomenclature	Root
1	methane
2	ethane
3	propane
4	butane
5	pentane
6	hexane
7	heptane
8	octane
9	nonane
10	decane

Any organic compound containing a continuous chain of five carbon atoms has a name based upon the alkane pentane. In A and B above, the name is based upon propane because the longest carbon chain in both contains four carbon atoms.

EXAMINER'S TOP TIP

It is important to learn these roots as they are the basis of the naming system.

Functional groups

Having identified the longest carbon chain, it is necessary to identify the various functional groups and name them.

The table lists some of the common functional groups.

Examples

The carbon atoms in the longest chain are numbered. Each functional group is then named and added to the root. Examples are:

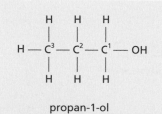

propan-1-ol

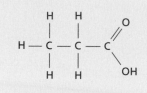

propanoic acid

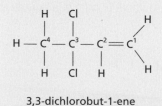

3,3-dichlorobut-1-ene

Where a name contains a number of prefixes the prefixes are arranged in alphabetical order. In 3-chloro-2,2-dimethylpropan-1-ol, the chloro prefix is placed before the methyl prefix (ignore di- and tri-).

EXAMINER'S TOP TIP
The chain is numbered to ensure that the lowest possible numbers appear in the name. Propan-1-ol could be named propan-3-ol if the numbering had been from the left-hand end.

Functional groups	Structure	Name as Prefix	suffix
double bond	$\diagdown C = C \diagup$	—	-ene
triple bond	$- C \equiv C -$	—	-yne
halogen (X = Cl, Br, I)	$-\overset{\mid}{\underset{\mid}{C}} - X$	chloro- bromo- iodo-	chloride bromide iodide
amine	$-\overset{\mid}{\underset{\mid}{C}} - N \overset{H}{\underset{H}{}}$	amino-	amine
hydroxyl	$-\overset{\mid}{\underset{\mid}{C}} - OH$	hydroxy-	-ol
carbonyl	$\diagdown C = O$	—	-al (aldehydes) -one (ketone)
carboxyl	$-C\overset{O}{\underset{O-H}{}}$	carboxy-	-oic acid
acid chloride	$-C\overset{O}{\underset{Cl}{}}$	—	-oyl chloride
amide	$-C\overset{O}{\underset{NH_2}{}}$	amido-	amide
acid anhydride	$-C\overset{O}{\underset{O}{}} - O - \overset{O}{\underset{C}{}}-$	—	-oic anhydride
ester	$-C\overset{O}{\underset{O-R}{}}$	—	-oate
nitrile	$-C \equiv N$	cyano-	nitrile

Quick test

1 Name the following compounds.

a $CH_3CH(OH)CH_2CH_3$ **b** $(CH_3)_3CBr$

c $CH_3CH_2CH_2CHO$ **d** $CH_3CH_2CH=CH_2$ **e** $HO_2CCH_2CH_2CO_2H$

Hydrocarbons – 1

Hydrocarbons are important compounds of carbon and hydrogen only.
There are millions of hydrocarbons because carbon atoms can form long chains and rings.
The bonds between carbon atoms are **covalent bonds**.
Hydrocarbons can be arranged in different homologous series.

What is a hydrocarbon?

A hydrocarbon is a compound of carbon and hydrogen **only**.
The displayed formulae of four hydrocarbons containing three carbon atoms are:

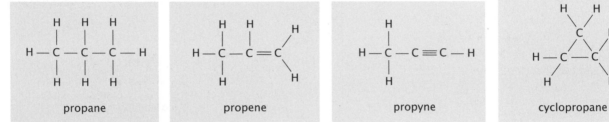

propane   propene   propyne   cyclopropane

Propane, propene and propyne are **aliphatic** or chain compounds.
Cyclopropane is a **ring compound**.
Propane and cyclopropane are called **saturated hydrocarbons**. This is because they contain only single carbon–carbon covalent bonds.
Propene and propyne are called **unsaturated hydrocarbons**. They contain a multiple bond (i.e. a double or triple bond).

The four compounds come from different **homologous series** (families):

> propane is an **alkane**
>
> propene is an **alkene**
>
> propyne is an **alkyne**
>
> cyclopropane is a **cycloalkane**.

Compounds in the same homologous series:
- differ from one another by $-CH_2-$
- have the same general formula
- contain the same functional group
- have similar chemical properties.

EXAMINER'S TOP TIP
The molecular formula of propane is C_3H_8. The structural formula is $CH_3CH_2CH_3$. The displayed formula is shown above. Look at what the question requires. Sometimes a displayed formula is essential to make it clear what is happening.

Sources of hydrocarbons

Most hydrocarbons are obtained from **crude oil (petroleum)**. Crude oil is a **fossil fuel** formed by the action of high temperatures and high pressures on dead sea animals over millions of years.
Crude oil as it comes from the earth is of little use. It has to be refined by the process of **fractional distillation**.

Fractional distillation of crude oil
Crude oil vapour passes into a tall column. As the gas passes up the column, compounds condense at different levels depending upon the boiling point. Compounds with low boiling points condense near the top of the column.
At different levels in the column **different fractions** are tapped off. Each fraction consists of a mixture of hydrocarbons boiling with a temperature range. Each fraction has different uses.

The process is summarised in the diagram.

Cracking

Oil companies can sell some fractions, e.g. gasoline (petrol), easily but have more difficulties selling higher boiling point fractions. **Cracking** is a process of breaking down long-chain saturated hydrocarbons into smaller hydrocarbons. This is in order to:

- produce more short-chain hydrocarbons suitable for petrol
- produce chemicals used to make other organic compounds including ethanol (page 82) and polymers (page 79).

Thermal cracking

The naphtha fraction is mixed with steam and heated under high pressure to a high temperature. A mixture of straight chain alkanes and alkenes is formed with a small amount of hydrogen.

e.g. $C_{10}H_{22} \rightarrow C_8H_{18} + C_2H_4$

Catalytic cracking

The vapour of heavy long-chain fractions is passed over a zeolite catalyst at 450 °C. This process produces more branched and ring hydrocarbons than thermal cracking. These are used in petrol.

Reforming

Heat and pressure is used to convert unbranched molecules into ring compounds. The products are used in petrol. e.g.

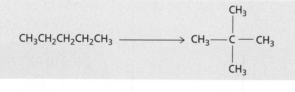

Isomerism

Isomerism converts unbranched hydrocarbons into branched hydrocarbons needed for petrol. e.g.

Diagram (right column)

refinery gas (calor gas, camping gas) $C_1 - C_4$

25 °C

gasoline (petrol) $C_4 - C_{10}$

naphtha (petrochemicals) $C_7 - C_{14}$

kerosine (aviation fuel) $C_{11} - C_{15}$

200 °C

gas oil (diesel, oil central heating) $C_{15} - C_{19}$

mineral oil (lubricating oil) $C_{20} - C_{30}$

crude oil

heater

400 °C

residue – a complex mixture requiring further refining:
· fuel oil (fuel for ships & power stations) $C_{30} - C_{40}$
· wax, grease (candles, grease, polish) $C_{41} - C_{50}$
· bitumen (road surfacing, roofing) > C_{50}

Reforming reaction:

$CH_3CH_2CH_2CH_2CH_2CH_2CH_3 \xrightarrow[\text{heat, pressure}]{\text{Pt catalyst}}$ methylbenzene (ring with CH_3) + $4H_2$

Isomerism reaction:

$CH_3CH_2CH_2CH_2CH_3 \longrightarrow$ branched structure with central C bonded to three CH_3 and one CH_3

Quick test

1. Each homologous series has a general formula which fits all members.
   Three of these general formulae are C_nH_{2n-2}, C_nH_{2n} and C_nH_{2n+2}.
   Which homologous series fits each formula?

2. Thermal cracking of the alkane containing 12 carbon atoms can produce ethene and ethane.
   Write a symbol equation for this reaction.

1. C_nH_{2n-2} for alkynes, C_nH_{2n} for alkenes and cycloalkanes, C_nH_{2n+2} for alkanes. 2. $C_{12}H_{26} \rightarrow 5C_2H_4 + C_2H_6$

Hydrocarbons – 2

All hydrocarbons burn in air and many are used as fuels.

Alkanes and cycloalkanes are generally unreactive.

Alkenes and alkynes have a wide range of reactions often involving addition.

Combustion of hydrocarbons

All hydrocarbons burn in excess air to produce **carbon dioxide** and **water**.

e.g. methane $CH_4 + 2O_2 \rightarrow CO_2 + 2H_2O$
      ethene $C_2H_4 + 3O_2 \rightarrow 2CO_2 + 2H_2O$

Combustion of hydrocarbons can lead to certain environmental problems.

- Incomplete combustion of hydrocarbons leads to the production of **carbon monoxide**. This is highly toxic and leads to about 50 deaths each year in Great Britain, usually due to gas appliances not being regularly serviced.
- Burning hydrocarbon fuels also produces nitrogen oxides and unburnt hydrocarbons in the atmosphere. These can be reduced by fitting catalytic converters to vehicles.
- Hydrocarbons from crude oil contain sulphur compounds. These are removed before hydrocarbon fuels are burned. Otherwise, sulphur dioxide would be produced and this would lead to acid rain.
- Carbon dioxide produced on the combustion of hydrocarbons contributes to the Greenhouse Effect.

Alkanes

Alkanes are generally unreactive. Apart from burning, the only common reactions involve reactions with the halogens.

Substitution reactions of alkanes

Alkanes react with halogens in the presence of ultraviolet radiation to form **halogenoalkanes**.

For example, methane reacts with chlorine in the presence of ultraviolet radiation to form a series of products.

$CH_4 + Cl_2 \rightarrow CH_3Cl + HCl$
chloromethane

$CH_3Cl + Cl_2 \rightarrow CH_2Cl_2 + HCl$
dichloromethane

$CH_2Cl_2 + Cl_2 \rightarrow CHCl_3 + HCl$
trichloromethane

$CHCl_3 + Cl_2 \rightarrow CCl_4 + HCl$
tetrachloromethane

EXAMINER'S TOP TIP
These reactions are substitution reactions. A halogen atom substitutes for a hydrogen atom in the alkane molecule. This can be repeated until all the hydrogen atoms have been replaced.

Ultraviolet radiation produces **free radicals** and a **chain reaction** is set up (see page 85).

Alkenes

Alkenes contain a carbon–carbon double bond.
e.g.

ethene

Between the two carbon atoms there are four bonding electrons. This high electron density makes alkenes more reactive than alkanes.

Addition of bromine
The addition reaction of alkene and bromine is used as a test for unsaturation.

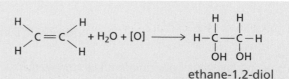

e.g.   unsaturated   saturated

ethene   1,2-dibromoethane
orange colour   colourless

Addition using acidified manganate(VII)
Cold, dilute acidified manganate(VII) turns from purple to colourless. This is used as a test for unsaturation
Ethane-1,2-diol is used as antifreeze in car engines.

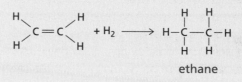

ethane-1,2-diol

Addition reaction with hydrogen
When a mixture of ethene and hydrogen is passed over a nickel catalyst at 140 °C an addition reaction takes place. This reaction is a **reduction** reaction and it is sometimes called **hydrogenation**. Hydrogenation of animal or vegetable fats and oils (that contain unsaturated compounds) produces margarine. The animal or vegetable fats and oils are liquid but after hydrogenation the product is solid.

ethane

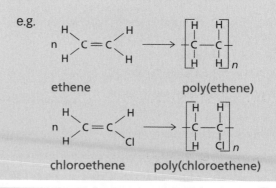

e.g.

ethene   poly(ethene)

Polymerisation of alkenes
Alkene molecules (called **monomers**) can join together by a series of addition reactions to produce a long chain **polymer**. Different polymers are produced by using different alkenes.

chloroethene   poly(chloroethene)

Quick test

1 Propane is used in camping gas cylinders.
Write a symbol equation for the burning of propane in excess oxygen.

2 Write a symbol equation for the combustion of methane in a limited supply of air.

3 Substituting two hydrogen atoms with chlorine atoms in ethane can produce two different isomers.
Draw the displayed formulae of the two isomers.

4 Hydrogenation of ethyne, C_2H_2, can produce two products. What are these two products?

5 Addition of hydrogen bromide, HBr, to propene can produce two products.
Write down the names and displayed formulae of these two compounds.

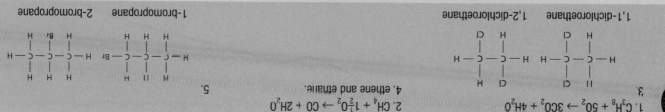

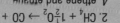

1. $C_3H_8 + 5O_2 \rightarrow 3CO_2 + 4H_2O$
2. $CH_4 + 1\frac{1}{2}O_2 \rightarrow CO + 2H_2O$
4. ethene and ethane.

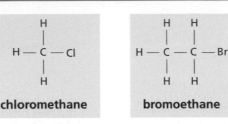

Halogenoalkanes

Halogenoalkanes are formed when a hydrogen atom in an alkane is replaced by a halogen atom.

Halogenoalkanes react to form a wide range of different compounds and this makes them useful for synthesising compounds.

Examples of halogenoalkanes

Examples of halogenoalkanes are:

chloromethane   bromoethane

Carbon and halogens have different electronegativities. Halogenoalkanes are polar molecules.

Comparative reactivity of halogenoalkanes

Chlorine is more electronegative than carbon. The electrons in the C–Cl bond move slightly towards the chlorine atom creating $\delta+$ and $\delta-$ charges. Polarity decreases down the halogen group.

$C^{\delta+} F^{\delta-}$
$C^{\delta+} Cl^{\delta-}$ Polarity decreases
$C^{\delta+} Br^-$
$C^{\delta+} I^{\delta-}$

This order would suggest that fluoroalkanes will react faster than chloroalkanes which react faster than bromoalkanes which react faster than iodoalkanes.

Bond enthalpy of C–X bond

The table gives the bond enthalpies of different C–X bonds.

Bond	Bond enthalpy/kJ mol^{-1}
C–F	467
C–Cl	364
C–Br	290
C–I	228

This order suggest that iodoalkanes will react faster than bromoalkanes which will react faster than chloroalkanes which react faster than fluoroalkanes.

Electronegativity and bond energy are in competition.

Because, in practice, reactivity is greater with iodoalkanes, the differences in bond enthalpy must be more significant.

In fact, iodoalkanes react more rapidly than bromoalkanes which react more rapidly than chloroalkanes.

Reactions of halogenoalkanes

Halogenoalkanes react readily with a variety of reagents to undergo nucleophilic substitution reactions (page 87). Some of the common reactions include the following.

Reaction with potassium hydroxide

Potassium hydroxide in aqueous solution reacts with halogenoalkanes on refluxing to produce an alcohol. These reactions are **substitution** reactions:

e.g. $CH_3CH_2Br + OH^- \rightarrow CH_3CH_2OH + Br^-$
          ethanol

A different reaction takes place if a halogenoalkane is refluxed with a solution of potassium hydroxide dissolved in ethanol. The reaction is an **elimination** reaction and produces an **alkene**:

e.g. $CH_3CH_2Br + OH^- \rightarrow CH_2{=}CH_2 + H_2O + Br^-$
          ethene

Reaction of a halogenoalkane with potassium cyanide

A halogenoalkane is convened to a **nitrile** (or cyanide) by refluxing the halogenoalkane with a solution of potassium cyanide in ethanol.

e.g. $CH_3CH_2Br + CN^- \rightarrow CH_3CH_2CN + Br^-$
          propanenitrile

Reaction of a halogenoalkane with ammonia

When a solution of a halogenoalkane in ethanol is heated with ammonia in a sealed vessel an **amine** is produced, but the reaction does not stop there.

e.g. $CH_3CH_2Br + NH_3 \rightarrow CH_3CH_2NH_2 + HBr$
          ethylamine

$CH_3CH_2Br + CH_3CH_2NH_2 \rightarrow (CH_3CH_2)_2NH + HBr$
          diethylamine

$CH_3CH_2Br + (CH_3CH_2)_2NH \rightarrow (CH_3CH_2)_3N + HBr$
          triethylamine

$CH_3CH_2Br + (CH_3CH_2)_3N \rightarrow (CH_3CH_2)_4N^+Br^-$
          tetraethylammonium bromide

EXAMINER'S TOP TIP
Reactions of halogenoalkanes with water, dilute acids, or alkalis are hydrolysis reactions. They involve water breaking the C–X bond, where X is a halogen.

EXAMINER'S TOP TIP
This reaction is an important step in increasing the number of carbon atoms in a molecule (called ascending the homologous series).

EXAMINER'S TOP TIP
Halogenoalkanes react very rapidly with water. Any other substitution reaction needs to use a solvent other than water if it is to take place. Ethanol is a suitable solvent.

Quick test

1 Why does water react with bromoethane but not with ethane?

2 Write down the products when iodoethane reacts with

  a an aqueous solution of potassium hydroxide

  b a solution of potassium hydroxide dissolved in ethanol.

3 What types of reaction are taking place in Q2.

4 Why is only one product possible when iodomethane reacts with potassium hydroxide?

5 Bromoethane can be prepared by an addition reaction. Suggest two substances that react to form bromoethane.

5. ethene and hydrogen bromide
4. Only methanol is produced. There is only one carbon atom so elimination cannot take place. This involves the formation of a double bond between two carbon atoms.
3. (a) (Nucleophilic) substitution; (b) elimination
2. (a) ethanol and potassium bromide; (b) ethene, potassium bromide and water.
1. Bromoethane contains a polar C–Br bond but ethane contains only C–H bonds which are not polar.

Alcohols

Alcohols are compounds containing the **–OH or hydroxyl group**.
Ethanol, commonly called just **'alcohol'**, is one of the homologous series of alcohols.
Examples of simple alcohols include

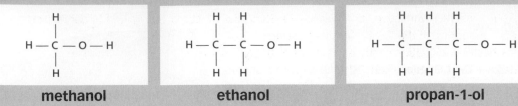

| methanol | ethanol | propan-1-ol |

Primary, secondary and tertiary alcohols

The classification depends upon the position of the OH group in the molecule.
In a primary alcohol, the OH group is attached to a carbon atom which is in turn attached directly to two or three hydrogen atoms.

It can be represented as:

$$R - \overset{\displaystyle H}{\underset{\displaystyle H}{C}} - OH$$

Only one hydrogen is attached to this carbon atom in a secondary alcohol:

$$R - \overset{\displaystyle H}{\underset{\displaystyle R}{C}} - OH$$

In a tertiary alcohol there are no hydrogen atoms attached to this carbon atom:

$$R - \overset{\displaystyle R}{\underset{\displaystyle R}{C}} - OH$$

> **EXAMINER'S TOP TIP**
>
> Make sure you understand the differences between primary, secondary and tertiary alcohols.

The four alcohols with a molecular formula C_4H_9OH have the following displayed formulae.

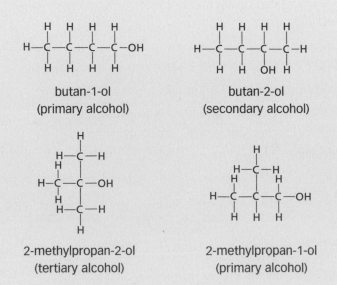

butan-1-ol
(primary alcohol)

butan-2-ol
(secondary alcohol)

2-methylpropan-2-ol
(tertiary alcohol)

2-methylpropan-1-ol
(primary alcohol)

Manufacture of ethanol

There are different methods of preparing ethanol:

1 Ethanol, C_2H_5OH, can be prepared by fermentation of starch or sugar solution in the presence of enzymes in yeast in anaerobic conditions (i.e. in absence of air), as in the wine-making process, e.g.

$$C_6H_{12}O_6 \ (aq) \rightarrow 2C_2H_5OH(aq) + 2CO_2(g)$$

This is a batch process.

This process is carried out in slightly warm conditions (about 30–35 °C). Air must not be allowed to come into contact with the solution, or souring of the wine might take place.

The souring involves the bacterial oxidation of the alcohol to a carboxylic acid.

$$C_2H_5OH(aq) + 2[O] \rightarrow CH_3COOH(aq) + H_2O(l)$$

The solution resulting from anaerobic fermentation is a dilute solution of ethanol in water. This solution can be concentrated by fractional distillation.

2 Ethene and steam are passed over a phosphoric acid catalyst at 330 °C under high pressure. (60–80 atmospheres).

This is a continuous process.

$$C_2H_4(g) + H_2O(g) \rightarrow C_2H_5OH(l)$$

The industrial preparation used for ethanol may depend on the raw material available.

A country where crude oil is refined may produce ethene and then use this to produce ethanol.

In a country without oil reserves and oil refineries (especially with hot climates), sugar may provide a renewable source for ethanol production.

Oxidation of alcohols

Oxidation of alcohols is used to distinguish primary, secondary and tertiary alcohols.

Primary alcohols are oxidised to produce an **aldehyde** and then, on prolonged oxidation, a **carboxylic acid**.

Secondary alcohols are oxidised to produce a **ketone**. No further oxidation takes place under the stated conditions.

Tertiary alcohols are only oxidised under very severe conditions, when the molecule is split. Under the conditions stated below, no oxidation takes place.

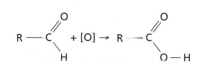

aldehyde   carboxylic acid

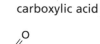

ketone

Methods for oxidising alcohols

- Heat the alcohol with a solution of potassium dichromate(VI) acidified with dilute sulphuric acid. During the oxidation the orange solution turns green owing to the formation of $Cr^{3+}(aq)$ ions. The aldehyde usually distils off during the oxidation as it has a lower boiling point than the corresponding alcohol.
- Heat the alcohol with either an acidic or an alkaline solution of potassium manganate(VII). The purple colour is removed during the oxidation. A brown precipitate of manganese(IV) oxide may be formed if an alkaline solution is used.
- Passing the alcohol vapour over a heated copper catalyst,

    e.g. $CH_3CH_2OH \rightarrow CH_3CHO + H_2$

Quick test

1 Draw a graphical formula for a secondary alcohol containing three carbon atoms.

2 Draw a graphical formula for a tertiary alcohol containing five carbon atoms.

3 Complete the following sentences.

    Oxidation of a primary alcohol produces first an _____ and on further oxidation a _____. Oxidation of a secondary alcohol produces a _____.

3. aldehyde, carboxylic acid, ketone

Organic mechanisms – 1

Understanding how organic reactions take place is called organic mechanisms.

The key to understanding <u>reaction mechanisms</u> is understanding <u>how a covalent bond can be broken</u>.

<u>Free radicals</u> are very <u>reactive atoms</u> or groups of atoms containing a <u>single electron</u>.

Breaking a covalent bond

Breaking a covalent bond is called **fission**.
There are two ways of breaking a covalent bond.
Consider the breaking of the C–C bond in ethane.

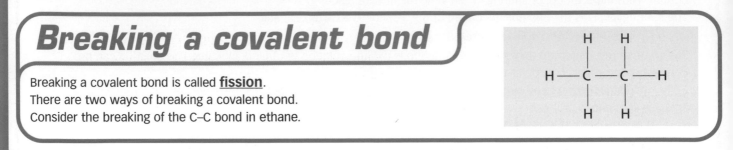

Homolytic fission

The bond is broken so that, one of the two electrons in the C–C bond goes to each carbon atom. The species formed are methyl **free radicals** and they each contain a single unpaired electron.

Free radicals are extremely short-lived and readily undergo further reaction. This type of bond breaking usually occurs in the gas phase or in non-polar solvents and in ultraviolet light.

> **EXAMINER'S TOP TIP**
> Homolytic fission is a fair way of breaking the covalent bond and heterolytic fission is an unfair way.

Heterolytic fission

The bond is broken so that one carbon atom receives both electrons while the other receives none.

The positively charged ion is called a **carbocation** (carbonium ion). The positively charged carbon atom is liable to attack by negatively charged species called **nucleophiles**.

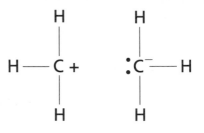

> **EXAMINER'S TOP TIP**
> Remember nucleophile and negative both begin with the letter 'n'.

The negatively charged ion is called a **carbanion**. It is liable to attack by positively charged species called **electrophiles**.
These kinds of ion are likely to exist in polar solvents, e.g. water.

Although free radicals, carbocations and carbanions may only exist for the shortest possible period of time, they are important in controlling how a reaction takes place.

Stability of carbocations

There are four possible carbocations with a formula $C_4H_9^+$.
They are:

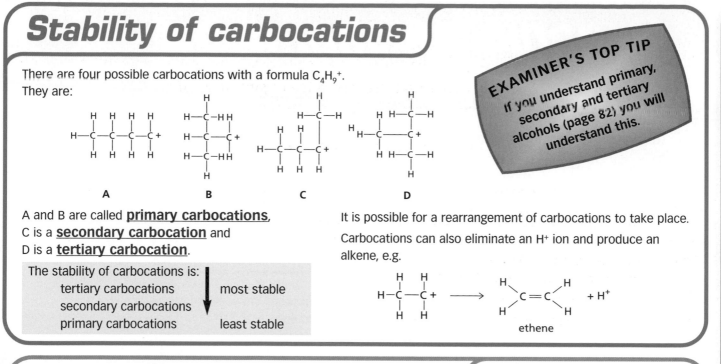

A B C D

A and B are called **primary carbocations**,
C is a **secondary carbocation** and
D is a **tertiary carbocation**.

The stability of carbocations is:
tertiary carbocations — most stable
secondary carbocations
primary carbocations — least stable

EXAMINER'S TOP TIP

If you understand primary, secondary and tertiary alcohols (page 82) you will understand this.

It is possible for a rearrangement of carbocations to take place.

Carbocations can also eliminate an H^+ ion and produce an alkene, e.g.

ethene

Free radical chain reaction

A most commonly used example of a free radical chain reaction is the reaction between methane and chlorine.
This reaction takes place when a mixture of the gases methane and chlorine are subjected to ultraviolet radiation.
The uv radiation reacts with a few chlorine–chlorine bonds and forms chlorine free radicals (chlorine atoms).

$$Cl_2 \rightarrow Cl\bullet + Cl\bullet \qquad \text{Initiation stage}$$

The following steps then take place.

$$Cl\bullet + CH_4 \rightarrow CH_3\bullet + HCl \quad \text{Propagation stages}$$
$$CH_3\bullet + Cl_2 \rightarrow CH_3Cl + Cl\bullet$$

The following reactions also take place but do not give further reaction.

$$CH_3\bullet + Cl\bullet \rightarrow CH_3Cl \qquad \text{Termination stage}$$
$$Cl\bullet + Cl\bullet \rightarrow Cl_2$$
$$CH_3\bullet + CH_3\bullet \rightarrow C_2H_6$$

EXAMINER'S TOP TIP

In the initiation stage some free radicals are formed. In the propagation each reaction of free radicals produces more free radicals and so continues the reaction. This is called a <u>chain reaction</u>. In the termination stages two free radicals react to produce a product that is not a free radical.

The mechanism explains why, in practice, small amounts of ethane are detected in the products.

Quick test

1 What would be formed if a chlorine molecule, Cl_2, was broken into two fragments by heterolytic fission?

2 Hydrogen and chlorine react explosively in a free radical chain reaction.

  a Suggest conditions for this reaction.

  b Write down a symbol equation for the whole reaction.

  c Write down an equation for the initiation stage producing chlorine free radicals.

  d Write down equations for the propagation stage.

  e Write down equations for the termination stage.

1. Cl+ and Cl⁻   2. (a) mixture of gases in sunlight or uv radiation;   (b) H₂(g) + Cl₂(g) → 2HCl(g);   (c) Cl₂ → Cl• + 2Cl•
(d) Cl• + H₂ → HCl + H•; H• + Cl₂ → HCl + Cl•;   (e) H• + H• → H₂; H• + Cl• → HCl; Cl• + Cl• → Cl₂

85

Organic mechanisms – 2

- Two types of organic reaction are addition and substitution.
- The reaction of alkenes where the double bond becomes a single bond (e.g. the addition of bromine to ethene) is an <u>electrophilic addition</u>.
- The reaction of halogenoalkanes to form alcohols are examples of <u>nucleophilic substitution</u> reactions.

Addition to alkenes (electrophilic addition)

In the carbon-carbon double bond there are four electrons in the double bond between the two carbon atoms.
There is a concentration of negative charge between the two carbon atoms.
This makes it a target for attack by electrophiles (positive species).

Ethene and HBr

There is a dipole within the hydrogen bromide molecule which gives the hydrogen atom a slight positive charge (shown by $\delta+$) and the bromine atom a slight negative charge (shown by $\delta-$).

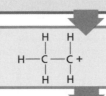

1 The first step involves the formation of a weak complex between the positive end of the HBr molecule and the electrons of the carbon=carbon double bond.

2 This weak complex may then convert into a more stable carbocation.

3 This then reacts rapidly with a bromide ion to form the product.

Ethene and Br_2

The bromine molecule Br–Br does not contain a permanent dipole.
An induced dipole is formed when it approaches an ethene molecule.

Propene and HBr

The reaction between propene and hydrogen bromide can, in theory, lead to two products.

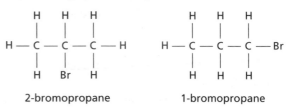

2-bromopropane   1-bromopropane

This can occur because the groups attached to the two carbon atoms, joined by the double bond, are different. The alkene is not symmetrical. In practice, addition produces 2-bromopropane only, according to **Markownikoff's rule**.

Markownikoff's rule states that during addition the more negative part of the molecule adding to the alkene (Br in this case) adds to the carbon atom attached to the lesser number of hydrogen atoms.

This rule can be explained in terms of stability of carbocations.
There are two carbocations:

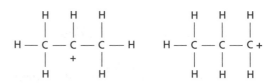

The secondary carbocation is more stable than the primary. It is formed in preference and leads to 2-bromopropane.

Substitution in halogenoalkanes (nucleophilic substitution)

For example $CH_3CH_2Br + OH^- \rightarrow CH_3CH_2OH + Br^-$

The carbon atom attached to the halogen is attacked by the nucleophile OH^-.
The reaction of a halogenoalkane with aqueous hydroxide ions can occur in two ways.

S_N1 mechanism

This is a two stage process that first involves the loss of an X^- ion and the formation of a carbocation. The carbocation ion reacts rapidly with an OH^- ion.

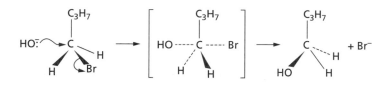

EXAMINER'S TOP TIP
Nucleophiles are atoms or groups of atoms where there is either a negative charge, e.g. CN^-, or a lone pair of non-bonding electrons, e.g. NH_3

S_N2 mechanism

This is a one-stage process involving a simultaneous loss of X^- and a gain of OH^- ion.

Elimination reactions occur when halogenoalkanes are refluxed with sodium hydroxide dissolved in ethanol.

When a halogenoalkane is heated with aqueous alkali, both substitution and elimination are possible. Substitution is so much more favourable so that elimination does not, in practice, occur. In the reaction of a halogenoalkane with a solution of alkali in ethanol, the substitution does not occur, and so elimination takes place.

When the alkali is dissolved in ethanol, the hydroxide ions are solvated with ethanol molecules. The resulting species is too large to approach the positive centre in the carbocation. Instead the hydroxide ion acts as a base and removes a proton.

EXAMINER'S TOP TIP
This is an example of where different conditions can lead the same chemicals to produce different products. At AS level it is important to give the conditions of chemical reactions.

Quick test

1 From the list select examples of electrophiles and nucleophiles.

CN^- NO_2^+ NH_3 Cl Br^+

2 But-1-ene reacts with hydrogen bromide by addition.

a Is this electrophilic, free radical or nucleophilic?

b Write down the displayed formulae of the two possible products.

c Which product is formed according to Markownikoff's rule.

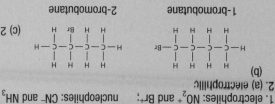

Exam-style questions
Use the questions to test your progress. Check your answers on pages 92–93.

Organic chemistry

Use data from the Data section (pages 90–91) to help you answer these questions.

1 A sample of a gaseous organic compound X contains 38.7% carbon, 16.1% hydrogen and 45.2% nitrogen.
  a How do you know that X does not contain oxygen? [1]

....

  b Calculate the empirical formula of X. [2]

  c 0.129 g of X occupies 100 cm³ at room temperature and atmospheric pressure. Calculate the relative molecular mass of X (1 mole of a gas occupies 24 dm³ at room temperature and atmospheric pressure). [2]

  d What is the molecular formula of X? [1]

2 Four compounds have the same empirical formula of CH_2. The compounds all have the same relative formula mass, i.e. 56 g.
  a What name is given to compounds with the same empirical formula and the same relative formula masses but different structural formulae? [1]

....

  b What is the molecular formula of each compound? [1]

....

  c Draw possible structures of these four compounds. (Hint: one is a ring compound.) [2]

  d Describe how the ring compound could be identified using bromine. [3]

....

3 Propene, C_3H_6, is an alkene.

  a Draw the displayed formula of propene. [1]

  b Propene reacts with bromine.
  (i) Draw the displayed formula and write the name of the product. [2]

....

  (ii) Underline the correct answer below.
   The mechanism of this reaction is: electrophilic addition

                                          nucleophilic addition

                                          free radical addition [1]

  c Propene undergoes addition polymerisation. Draw the displayed formula of poly(propene). [1]

d Draw the displayed formula and name the compound formed when propene reacts with hydrogen bromide. [2]

4 Epoxyethane is a compound with the displayed formula

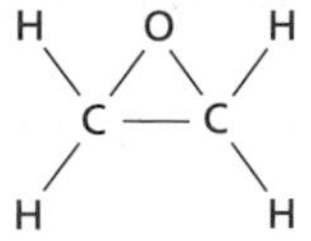

a Write down the molecular formula of epoxyethane. [1]

....

b Epoxyethane is produced by the reaction of ethene with oxygen in the presence of a silver-based catalyst. Write an equation for the reaction. [2]

....

c Ethane-1,2-diol is produced when epoxyethane reacts with water in the presence of dilute sulphuric acid at 60 °C.
(i) Write an equation for this reaction. (ii) Suggest one use for ethane-1,2-diol. [2]

....

5 Three alcohols are shown below.

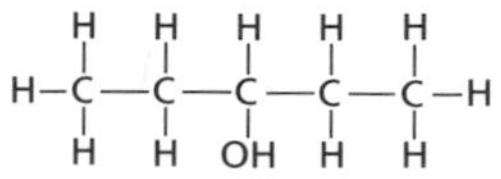

a Classify each alcohol as primary, secondary or tertiary. [3]

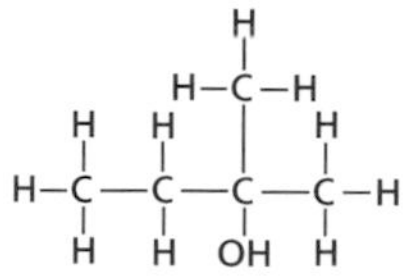

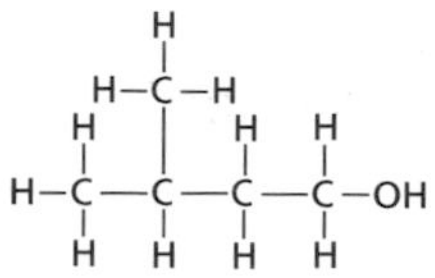

....

b Write systematic names for A, B and C. [3]

....

c Describe how these three alcohols could be distinguished by chemical tests using potassium dichromate(VI). [3]

....

....

d Dehydration of these alcohols involves removing water. Explain why dehydration of A and C each produce a single product but dehydration of B can produce two products. [3]

6 a Draw the displayed formula of 1-iodopropane. [1]

b (i) Draw the displayed formula of the carbocation formed when the C–I bond is broken. [1]

(ii) Is this carbocation primary, secondary or tertiary? [1]

....

c What type of bond fission has taken place? [1]

....

d Draw the displayed formula of another carbocation with the same formula. [1]

Total: / 41

Data

Periodic Table

Key

relative atomic mass	→	1.0
atomic symbol	→	H
atomic number	→	1
name	→	Hydrogen

Group 1	Group 2											Group 3	Group 4	Group 5	Group 6	Group 7	Group 0
1.0 **H** 1 Hydrogen																	4.0 **He** 2 Helium
6.9 **Li** 3 Lithium	9.0 **Be** 4 Beryllium											10.8 **B** 5 Boron	12.0 **C** 6 Carbon	14.0 **N** 7 Nitrogen	16.0 **O** 8 Oxygen	19.0 **F** 9 Fluorine	20.2 **Ne** 10 Neon
23.0 **Na** 11 Sodium	24.3 **Mg** 12 Magnesium											27.0 **Al** 13 Aluminium	28.1 **Si** 14 Silicon	31.0 **P** 15 Phosphorus	32.1 **S** 16 Sulphur	35.5 **Cl** 17 Chlorine	39.9 **Ar** 18 Argon
39.1 **K** 19 Potassium	40.1 **Ca** 20 Calcium	45.0 **Sc** 21 Scandium	47.9 **Ti** 22 Titanium	50.9 **V** 23 Vanadium	52.0 **Cr** 24 Chromium	54.9 **Mn** 25 Manganese	55.8 **Fe** 26 Iron	58.9 **Co** 27 Cobalt	58.7 **Ni** 28 Nickel	63.5 **Cu** 29 Copper	65.4 **Zn** 30 Zinc	69.7 **Ga** 31 Gallium	72.6 **Ge** 32 Germanium	74.9 **As** 33 Arsenic	79.0 **Se** 34 Selenium	79.9 **Br** 35 Bromine	83.8 **Kr** 36 Krypton
85.5 **Rb** 37 Rubidium	87.6 **Sr** 38 Strontium	88.9 **Y** 39 Yttrium	91.2 **Zr** 40 Zirconium	92.9 **Nb** 41 Niobium	95.9 **Mo** 42 Molybdenum	– **Tc** 43 Technetium	101 **Ru** 44 Ruthenium	103 **Rh** 45 Rhodium	106 **Pd** 46 Palladium	108 **Ag** 47 Silver	112 **Cd** 48 Cadmium	115 **In** 49 Indium	119 **Sn** 50 Tin	122 **Sb** 51 Antimony	128 **Te** 52 Tellurium	127 **I** 53 Iodine	131 **Xe** 54 Xenon
133 **Cs** 55 Caesium	137 **Ba** 56 Barium	139 **La** 57 Lanthanum	178 **Hf** 72 Hafnium	181 **Ta** 73 Tantalum	184 **W** 74 Tungsten	186 **Re** 75 Rhenium	190 **Os** 76 Osmium	192 **Ir** 77 Iridium	195 **Pt** 78 Platinum	197 **Au** 79 Gold	201 **Hg** 80 Mercury	204 **Tl** 81 Thallium	207 **Pb** 82 Lead	209 **Bi** 83 Bismuth	– **Po** 84 Polonium	– **At** 85 Astatine	– **Rn** 86 Radon
– **Fr** 87 Francium	– **Ra** 88 Radium	– **Ac** 89 Actinium	– **Rf** 104 Rutherfordium	– **Db** 105 Dubnium	– **Sg** 106 Seaborgium	– **Bh** 107 Bohrium	– **Hs** 108 Hassium	– **Mt** 109 Meitnerium	– **Unn** 110 Ununnilium	– **Uuu** 111 Unununium	– **Uub** 112 Ununbium		– **Uuq** 114 Ununquadium		– **Uuh** 116 Ununhexium		– **Uuo** 118 Ununoctium

lanthanides

| 140 **Ce** 58 Cerium | 141 **Pr** 59 Praseodymium | 144 **Nd** 60 Neodymium | – **Pm** 61 Promethium | 150 **Sm** 62 Samarium | 152 **Eu** 63 Europium | 157 **Gd** 64 Gadolinium | 159 **Tb** 65 Terbium | 163 **Dy** 66 Dysprosium | 165 **Ho** 67 Holmium | 167 **Er** 68 Erbium | 169 **Tm** 69 Thulium | 173 **Yb** 70 Ytterbium | 175 **Lu** 71 Lutetium |

actinides

| – **Th** 90 Thorium | – **Pa** 91 Protactinium | – **U** 92 Uranium | – **Np** 93 Neptunium | – **Pu** 94 Plutonium | – **Am** 95 Americium | – **Cm** 96 Curium | – **Bk** 97 Berkelium | – **Cf** 98 Californium | – **Es** 99 Einsteinium | – **Fm** 100 Fermium | – **Md** 101 Mendelevium | – **No** 102 Nobelium | – **Lr** 103 Lawrencium |

Approximate relative atomic mass (to be used in calculations throughout the book)

H = 1; He = 4; C = 12; N = 14; O = 16; Na = 23; Mg = 24; Al = 27; P = 31; S = 32; Cl = 35.5; K = 39; Ca = 40; Fe = 56; Ni = 59; Cu = 64; Br = 80; Ag = 108; Sn = 118; Hg = 200; Pb = 207

Common half equations

Common oxidising agents

oxygen	$O_2(g) + 4e^- \rightarrow 2O^{2-}(aq)$
chlorine	$Cl_2(g) + 2e^- \rightarrow 2Cl^-(aq)$
bromine	$Br_2(g) + 2e^- \rightarrow 2Br^-(aq)$
iodine	$I_2(g) + 2e^- \rightarrow 2I^-(aq)$

manganate(VII) in acid solution

$MnO_4^-(aq) + 8H^+(aq) + 5e^- \rightarrow Mn^{2+}(aq) + 4H_2O(l)$

dichromate(VI) in acid solution

$Cr_2O_7^{2-}(aq) + 14H^+(aq) + 6e^- \rightarrow 2Cr^{3+}(aq) + 7H_2O(l)$

iron(III) salts

$Fe^{3+}(aq) + e^- \rightarrow Fe^{2+}(aq)$

hydrogen ions

$2H^+(aq) + 2e^- \rightarrow H_2(g)$

hydrogen peroxide in the absence of another oxidising agent

$H_2O_2(aq) + 2H^+(aq) + 2e^- \rightarrow 2H_2O(l)$

Common reducing agents

metals	$M(s) \rightarrow M^{n+}(aq) + ne^-$
	e.g. $Zn(s) \rightarrow Zn^{2+}(aq) + 2e^-$
iron(II) salts	$Fe^{2+}(aq) \rightarrow Fe^{3+}(aq) + e^-$

acidified potassium iodide

$2I^-(aq) \rightarrow I_2(g) + 2e^-$

thiosulphate	$2S_2O_3^{2-}(aq) \rightarrow S_4O_6^{2-}(aq) + 2e^-$
ethanedioates	$C_2O_4^{2-}(aq) \rightarrow 2CO_2(g) + 2e^-$
hydrogen	$H_2(g) \rightarrow 2H^+(aq) + 2e^-$

hydrogen peroxide in the presence of a strong oxidising agent $\quad H_2O_2(aq) \rightarrow O_2(g) + 2H^+(aq) + 2e^-$

Common bond enthalpies

bond	bond enthalpy /kJ mol^{-1}	bond	bond enthalpy /kJ mol^{-1}
H—H	436	C—H	412
C—C	348	N—H	388
C=C	612	O—H	463
C≡C	837	F—H	562
N≡N	944	Cl—H	431
F—F	158	Br—H	366
Cl—Cl	242	I—H	299
Br—Br	193	C—F	484
I—I	151	C—Cl	338
C—O	360	C—Br	276
C=O	743	C—I	238

Answers

Basic concepts (pp. 4–19)

1
 a $3Mn_3O_4 + 8Al \longrightarrow 9Mn + 4Al_2O_3$
 b $4NH_3 + 5O_2 \longrightarrow 4NO + 6H_2O$
 c $Na_2H_2P_2O_7 + 2NaHCO_3 \longrightarrow 2Na_2HPO_4 + 2CO_2 + H_2O$

2 $50\,cm^3$

3 0.5 moles of magnesium ions.
 0.6 moles of chloride ions combine with 0.3 moles of magnesium ions.
 0.2 moles of sulphate ions combine with 0.2 moles of magnesium ions.

4 $1000\,cm^3$ of sodium hydroxide solution ($0.1\,mol\,dm^{-3}$) contain 0.1 moles of sodium hydroxide.
 $25.0\,cm^3$ of $0.1\,mol\,dm^{-3}$ sodium hydroxide contain
 $\dfrac{0.1 \times 25.0\ moles}{1000} = 0.0025$ moles
 From the equation, 2 moles of sodium hydroxide react with 1 mole of sulphuric acid
 so 0.0025 moles of NaOH react with 0.00125 moles H_2SO_4
 0.00125 moles H_2SO_4 are present in $20.0\,cm^3$ sulphuric acid
 $\dfrac{0.00125 \times 1000}{20}$ moles H_2SO_4 would be present in $1000\,cm^3$
 Concentration of sulphuric acid = $0.0625\,mol\,dm^{-3}$

5
 a 272g
 b 0.735 tonnes
 c 0.471 tonnes
 d $176\,000\,dm^3$ (to 3 s.f.)

6
 a $Na_2CO_3(s) + Ca(OH)_2(aq) \longrightarrow CaCO_3(s) + 2NaOH(aq)$
 b $Mg(s) + 2AgNO_3(aq) \longrightarrow 2Ag(s) + Mg(NO_3)_2(aq)$
 c $2Cu(NO_3)_2(s) \longrightarrow 2CuO(s) + 4NO_2(g) + O_2(g)$
 d $2C_2H_6(g) + 7O_2(g) \longrightarrow 4CO_2(g) + 6H_2O(l)$

7 If the RAM of X is x, using the equation we can say that (x + 160)g of XBr_2 would form (x + 71)g of XCl_2.
 $\dfrac{1.500}{0.89} = \dfrac{x + 160}{x + 71}$ $x = 58.9$
 Using the Periodic Table, the element is cobalt, Co.

8 Number of moles $Na_2CO_3 = \dfrac{1.40}{106} = 0.0132$
 Molarity of $Na_2CO_3 = \dfrac{0.0132 \times 1000}{250} = 0.0528\,mol\,dm^{-3}$
 Molarity of HCl $= \dfrac{0.0528 \times 25.0 \times 2}{24.5 \times 1} = 0.108\,mol\,dm^{-3}$
 Concentration $= 0.108 \times 36.5 = 3.94\,g\,dm^{-3}$

9
 a Mass of oxygen in sample = 2.32 – 1.68 = 0.64 g
 Number of moles of iron = $\dfrac{1.68}{56} = 0.03$
 Number of moles of oxygen = $\dfrac{0.64}{16} = 0.04$
 Formula of the iron oxide is Fe_3O_4
 b $Fe_3O_4(s) + 4H_2(g) \longrightarrow 3Fe(s) + 4H_2O(l)$

Physical chemistry (pp. 22–45)

1
 a 56 protons, 56 electrons, 81 neutrons
 b 92 protons, 92 electrons, 143 neutrons
 c 92 protons, 92 electrons, 146 neutrons.

2 In 100 atoms, 60.2 will be gallium-69 and 39.8 will be gallium-71.
 Gallium-69 $60.2 \times 69 = 4153.8$
 Gallium-71 $39.8 \times 71 = 2825.8$
 Total mass = 6979.6
 Relative atomic mass = $\dfrac{6979.6}{100} = 69.8$

3 The three isotopes are hydrogen-1, hydrogen-2 and hydrogen-3.
 They all contain 1 proton and 1 electron.
 Hydrogen-1 contains no neutrons, hydrogen-2 contains 1 neutron and hydrogen-3 contains 2 neutrons.

4
 a $1s^2\,2s^2\,2p^6\,3s^2\,3p^3$
 b $1s^2\,2s^2\,2p^4$
 c $1s^2\,2s^2\,2p^6\,3s^2\,3p^6\,4s^2\,3d^2$
 d $1s^2\,2s^2\,2p^6\,3s^2\,3p^6\,4s^2\,3d^{10}\,4p^5$
 e $1s^2\,2s^2\,2p^6\,3s^2\,3p^6\,4s^2\,3d^{10}\,4p^3$

5
 a

species	protons	neutrons	electrons
sodium-23	11	12	11
oxygen-16	8	8	8
oxide-18 ion	8	10	10

 b (i) $1s^2\,2s^2\,2p^6\,3s^1$
 (ii) $1s^2\,2s^2\,2p^6$
 c Any three of the following:
 High melting point or high boiling point
 Conducts electricity when molten or in aqueous solution
 Soluble in water or soluble in polar solvents
 Colourless or white.
 d $44 = C^{16}O_2^+$, $46 = C^{16}O^{18}O^+$, $48 = C^{18}O_2^+$

6
 a RMM = 71 (the highest value on the diagram, corresponding to M^+)
 b Difference = 19 therefore atom is F
 c Difference = 38 therefore atoms are F_2
 d Difference = 57 therefore atoms are F_3
 e N^+ ion
 f NF_3

7 Trigonal planar e.g. BF_3 Tetrahedral e.g. CH_4
 Pyramidal e.g. NH_3 Bent e.g. H_2O

8 requires energy +412 +193 = +605 kJ gives out energy = –276 –366 = –642 kJ overall energy change = –642 +605 = –37 kJ mol^{-1}

Inorganic chemistry (pp. 48–67)

1
 a

Na	Mg	Al	Si	P	S	Cl	Ar
NaCl	$MgCl_2$	$AlCl_3$ or Al_2Cl_6	$SiCl_4$	PCl_3	SCl_2	Cl_2	—
				PCl_5	S_2Cl_2		

 b Example:
 Silicon(IV) chloride is prepared by passing dry chlorine gas over heated silicon.
 Silicon(IV) chloride is collected as a liquid by condensing the vapour using a U tube in cold water.
 c $NaCl(s) + aq \longrightarrow Na^+(aq) + Cl^-(aq)$
 $PCl_3(l) + 3H_2O(l) \longrightarrow H_3PO_3(aq) + 3HCl(aq)$

2
 a

Halogen	Physical state at room temperature	Colour
fluorine	gas	yellowish with greenish tinge
chlorine	gas	green/yellow
bromine	liquid	red/brown
iodine	solid	black

 b (i) $KCl(s) + H_2SO_4(conc) \longrightarrow KHSO_4(s) + HCl(g)$
 (ii) $HCl + H_2O \longrightarrow H_3O^+ + Cl^-$
  acid base   acid base
 Brønsted and Lowry defined acid as a proton donor and a base as a proton acceptor. Water accepts the proton and is a base.
 (iii) Silver nitrate solution contains silver ions and nitrate ions.
 When silver nitrate solution is added to a solution of Z (hydrochloric acid, containing chloride ions), a white precipitate of silver chloride is formed.
 $Ag^+(aq) + Cl^-(aq) \longrightarrow AgCl(s)$
 The silver chloride dissolves in concentrated ammonia solution.
 c $3ClO^-(aq) \longrightarrow 2Cl^-(aq) + ClO_3^-(aq)$
 O.S. +1, O.S. –1, O.S. +5
 Disproportionation occurs when the reactant is both oxidised (chlorate(I) to chlorate(V)) and reduced (chlorate(I) to chloride) in the same reaction.

3
 a Formula is Fe_3O_4
 b $Fe_2O_3(s) + 3CO(g) \longrightarrow 2Fe(l) + 3CO_2(g)$
 c Calcium carbonate decomposes to give calcium oxide and carbon dioxide. Calcium oxide reacts with silicon(IV) oxide to form calcium silicate (slag).
 This removes impurities from the furnace.
 d Add a suitable reagent e.g. dilute hydrochloric acid.
 Iron(III) oxide is basic.
 It forms a soluble compound: iron(III) chloride
 $Fe_2O_3(s) + 6HCl(aq) \longrightarrow 2FeCl_3(aq) + 3H_2O(l)$

4
 a The ionisation energy marked on the graph for K should be less than that for Na.
 b The number of shells increases down the group., therefore the outer electron is further from the nucleus.
 Less energy is required to remove it, therefore, ionisation energy decreases.
 c The pattern of 2 (Na, Mg), 3(Al, Si, P) 3 (S, Cl, Ar) is noted.
 The ionisation energy of Group 3 is lower than Group 2 because the electron is lost from the p orbital leaving a full stable s orbital. Also the p electron in Al is at a higher level than the s electrons in Na and Mg and is therefore, on average, further from the nucleus.

5 **a** Zinc is oxidised, H^+ in HCl is reduced,

    b One atom of copper in copper(I) chloride is oxidised to Cu^{2+} in $CuCl_2$ and one atom is reduced to Cu^0. This is disproportionation.

    c PbO_2 is reduced and HCl is oxidised.

    d Cr^{3+} in Cr_2O_3 is reduced to Cr^0. Al is oxidised to Al^{3+}.

    e Cu is oxidised, two NO_3^- groups are reduced.

Organic chemistry (pp. 70–87)

1 **a** Percentages of carbon, hydrogen and add up to 100%.

    b

$$\frac{38.7}{12} \quad \frac{16.1}{1} \quad \frac{45.2}{14}$$

$$3.2 \qquad 16.1 \qquad 3.2$$

$$1 \qquad\quad 5 \qquad\quad 1$$

$$CH_5N$$

    c $0.129 \times 240 = 30.96$

    d CH_5N

2 **a** Structural isomers

    b C_4H_8

    c

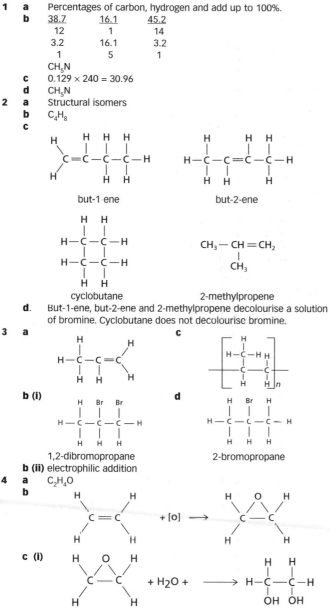

         but-1-ene   but-2-ene

         cyclobutane   2-methylpropene

    d. But-1-ene, but-2-ene and 2-methylpropene decolourise a solution of bromine. Cyclobutane does not decolourise bromine.

3 **a**   **c**

    b (i)   **d**

      1,2-dibromopropane   2-bromopropane

    b (ii) electrophilic addition

4 **a** C_2H_4O

    b

    c (i)

       (ii) antifreeze

5 **a** A is secondary; B is tertiary; C is primary.

    b A is pentan-3-ol; B is 2-methylbutan-2-ol; C is 3-methylbutan-1-ol.

    c Oxidation of A produces a ketone. Oxidation of B does not take place.

        Oxidation of C produces first an aldehyde and then a carboxylic acid.

    d The carbocation produced when A and C react can only form one product but with B there is a choice of product depending on which H^+ is removed.

6 **a**   **b (i)**

                                         (ii) primary

    c heterolytic fission

    d

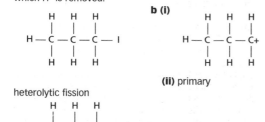

Glossary

Acid A substance which has a tendency to release protons (hydrogen ions). A strong acid, such as nitric acid, releases hydrogen ions readily. Weak acids, such as ethanoic acid, release hydrogen ions far less readily.

Activation energy The minimum energy required by particles in collision to bring about a chemical reaction.

Addition polymerisation The addition reaction of many molecules of monomer to form one large molecule of polymer.

Addition reaction A reaction in which two molecules react together to form one molecule.

Alcohols Alcohols have the general formula R-OH where R is an alkyl group. Their names end in -ol. Ethanol, C_2H_5OH is an alcohol.

Aldehyde A carbonyl compound with the general formula

The names of aldehydes end in -al. Ethanal, CH_3CHO is an aldehyde.

Alkali A base which dissolves in water, releasing hydroxide ions (OH^-).

Alkane A hydrocarbon with the general formula C_nH_{2n+2}.

Alkene A hydrocarbon with the general formula C_nH_{2n}.

Alkyl group A group of atoms with the general formula C_nH_{2n+1} forming part of a molecule.

Allotropes Different forms of the same element, having different structures. Carbon has several allotropes: diamond, graphite and the fullerenes.

Amphoteric Having both acidic and basic properties. For example, aluminium oxide is an amphoteric oxide. It forms salts both with acids and with alkalis.

Atom A single unit of an element.

Atomic mass see **Relative atomic mass**

Atomic number (symbol Z) The number of protons (and therefore electrons) present in an atom.

Atomic radius The distance from the centre of the nucleus to the outermost electrons in an atom.

Avogadro constant The number of particles contained in one mole of a substance: 6.022×10^{23} mol^{-1}.

Base A substance which has a tendency to gain protons. Bases which dissolve in water are called **alkalis**. Strong bases gain protons readily.

Bond energy (symbol E) The bond enthalpy: the amount of energy released when the bond is formed, which equals the amount of energy absorbed when the bond is broken.

Carbanion An anion where a carbon atom carries much of the negative charge. For example R_3C^-.

Carbocation (Carbonium ion) A positively charged group containing carbon and hydrogen atoms.

Carbonyl compounds Compounds containing the group >C=O.

Carboxylic acids Organic acids having the general formula RCOOH. Their names end in -oic acid. Ethanoic acid has the formula CH_3COOH.

Catalyst A substance that alters the rate of a chemical reaction by changing the activation energy. A catalyst remains chemically unchanged at the end of the reaction.

Closed system A system from which reactants and products cannot escape, and to which they cannot be added. Chemical equilibrium is only possible in a closed system.

Coordinate (dative) bond A covalent bond in which the shared electron pair originates from the same atom. It can be written as X→Y, showing that the shared electron pair comes from X.

Covalent bond A bond in which two atoms share one or more pairs of electrons. A hydrogen molecule, H–H, has a single covalent bond.

Cracking The process of breaking up the long-chain hydrocarbons in crude oil into shorter-chain hydrocarbons that can be used in the chemical industry and for petrol.

Crude oil A mixture of hydrocarbons formed naturally by the decomposition of marine animals over millions of years.

Dehydration reaction A reaction in which water is eliminated.

Delocalised electrons Electrons that are not located on one particular atom, but are free to move between all atoms in the structure. Delocalised electrons are found in metals and graphite.

Dipole Consists of a positive charge and an equal negative charge separated by a short distance. Temporary dipoles form in non-polar molecules, such as nitrogen. Some molecules, such as hydrogen chloride, are **polar**. They have permanent dipoles because the electron distribution is always uneven.

Dot-and-cross diagram A means of representing the electrons in a molecule. Electrons are drawn as either dots or crosses, to indicate their original atom.

Electron A negatively charged particle. Electrons orbit the atomic nucleus in energy levels. Atoms of different elements have different numbers of electrons.

Electronegativity The tendency of the atoms of an element to gain electrons. Elements whose atoms gain electrons easily are the most electronegative. Fluorine is the most electronegative element.

Electrophile An electron-seeking group. Electrophiles are positively charged. Examples include the nitryl group, $-NO_2^+$.

Electrophilic addition A reaction in which an electrophile is attracted to an area of high electron density. The electrophile adds on to the atom or group – an addition reaction.

Electrophilic substitution A reaction in which an electrophile is attracted to an area of high electron density. The electrophile replaces an atom or group – a substitution reaction.

Element A substance which cannot be broken down into any simpler substance by chemical means.

Elimination reaction A reaction in which a small molecule such as water is removed from a molecule to create a double bond.

Empirical formula The simplest formula of a compound, showing the ratios of the numbers of atoms in the molecule. For example, CH_3 is the empirical formula of ethane, C_2H_6.

Endothermic reaction A chemical reaction in which energy is absorbed.

Energy level One of the fixed range of energies to which electron energies in atoms are limited; sometimes described as **shells** and sub-shells. An electron in an atom requires a particular amount or quantum of energy to move from one energy level to the next.

Enthalpy Energy content.

Enthalpy change An amount of energy that is transferred (absorbed or released).

Enthalpy change of combustion The amount of energy transferred when one mole of a substance burns completely in oxygen under standard conditions.

Enthalpy change of vaporisation The amount of energy required to convert one mole of a liquid to a gas at its boiling point.

Equilibrium The state reached in a reversible reaction at which the rates of the two opposing reactions are equal, so that the system has no further tendency to change. This is a dynamic equilibrium, as reactants and products are both still being formed, but at equal rates.

Exothermic reaction A chemical reaction in which energy is released.

Fermentation The process by which, for example, the micro-organism yeast converts glucose into ethanol and carbon dioxide in order to release energy.

Fossil fuels Fuels that have been formed by the slow decomposition of plant and animal material. They include coal, oil, natural gas and peat.

Fractional distillation A separation method based on the difference in boiling points of substances, e.g. the process is used to separate the components of crude oil into groups of hydrocarbons of similar chain length. The process involves heating the crude oil until it vaporises and collecting the products within a set boiling point range.

Giant atomic structure A structure that contains many millions of atoms all bonded together. Diamond and graphite have giant atomic structures.

Half-equation Part of the equation for a redox reaction, showing the oxidation or reduction of one particular element. Two half-equations, showing the simultaneous oxidation and reduction steps, make up the full equation for the redox reaction.

Heterogeneous catalysis A reaction for which the catalyst and the reactants are in different phases.

Heterogeneous equilibrium An equilibrium in which the reactants are in different phases.

Heterolytic bond breaking The breaking of a single covalent bond (a bonding pair of electrons) so that the two electrons remain on one atom. Ions are formed.

Homogeneous catalysis A reaction for which the catalyst and the reactants are in the same phase.

Homogeneous equilibrium An equilibrium in which all the reactants are in the same phase.

Homologous series A series of organic compounds with the same general formula, each successive member of the series differs by a CH_2. For example, the alkanes form a homologous series, CH_4, C_2H_6, C_3H_8 etc.

Homolytic bond breaking The breaking of a single covalent bond (a bonding pair of electrons) so that one electron remains with each atom.

Hydrogen bonding The intermolecular bonding between dipoles in adjacent molecules in which hydrogen is bonded to a very electronegative element. For example, intermolecular hydrogen bonding exists in water (H_2O), ammonia (NH_3) and hydrogen fluoride (HF).

Hydrogenation The addition reaction of a compound with hydrogen.

Ideal gas A gas made up of particles of negligible size, with no forces acting between them.

Ideal gas equation A mathematical description of the relationship between volume, temperature and pressure. For an ideal gas: $pV = nRT$

Intermolecular bonding Bonding between molecules; the phrase does not refer to bonding within molecules. There are several types of intermolecular bonding, including van der Waal's forces and hydrogen bonds.

Intramolecular forces Forces between molecules. In carbon dioxide the covalent bonds between the atoms in each molecule (called intermolecular forces) are strong but the intramolecular forces between the molecules are weak.

Ion Particle, consisting of an atom or group of atoms, that carries a positive or negative electric charge. An atom forms an ion when it loses or gains one or more electrons.

Ionic equations A concise method of writing down the important changes that affect the ions directly involved in a chemical reaction.

Ionisation energy The energy required to remove one mole of electrons from one mole of atoms of an element so that the electrons are no longer under the influence of the positive charge of a nucleus.

Isomers Compounds that have the same molecular formula but different structural formulae.

Isotopes Atoms that have the same atomic number but different mass numbers.

Ketone A carbonyl compound with the general formula

$$R_2C=O$$

Propanone, CH_3COCH_3 is a ketone.

Lattice A geometrical arrangement of points. Crystal structures are based on lattices, with the particles positioned at the points.

Le Chatelier's principle When an equilibrium reaction mixture is subjected to a change in conditions, the composition of the mixture adjusts to counteract the change.

Mass number (symbol A) The number of protons plus the number of neutrons present in an atom.

Mass spectroscopy A technique used to find the relative atomic mass (symbol A_r) of an element or the relative molecular mass (symbol M_r) of a compound. It identifies the types and amounts of any isotopes present.

Molar gas constant (symbol R) The proportionality constant in the ideal gas equation.

Molar mass The mass of one mole of a substance. For example, the molar mass of magnesium is 24 g mol^{-1}.

Molar volume (symbol V_m) The volume occupied by one mole of any gas. It is 24.0 dm^3 at room temperature and atmospheric pressure.

Molarity The concentration of a solution expressed in mol dm^{-3}.

Mole An amount of substance that contains 6.022×10^{23} particles. These may be atoms, ions, molecules or electrons.

Molecular formula A formula showing the number and types of atoms present in a molecule.

Monomer A molecule that can react with many other similar molecules to build up a large molecule, or most polymer plastic materials are polymers; for example, poly(ethene) is a polymer of the monomer ethene.

Neutralisation The reaction of an acid and a base to form a salt and water.

Neutron A neutral (uncharged) mass particle found in the atomic nucleus. Its mass is approximately 1 atomic unit.

Nucleophile An atom or group of atoms that is attracted to a positive charge. NH_3, OH^- and H_2O can act as nucleophiles.

Nucleophilic substitution reaction A chemical reaction in which one nucleophile replaces another in a molecule.

Nucleus The central part of an atom, around which the electrons orbit. It consists of positively charged protons and neutral neutrons, tightly packed together.

Organic chemistry The chemistry of carbon compounds. Several million different carbon compounds are known.

Oxidation A process in which a species loses electrons. It can also be defined as an increase in oxidation state for an element. Oxidation and reduction occur together in a redox reaction.

Oxidation state The charge that an element would have if it were totally ionically bonded. For example, the oxidation state of hydrogen in water is +1, and that of oxygen in water is –2, even though water is covalently bonded. Oxidation state changes in a redox reaction.

Oxidising agent An element or compound that gains electrons from a reducing agent, which itself loses electrons in the process. The oxidising agent is reduced, and the reducing agent is oxidised.

Periodic Table A classification of the elements in order of increasing atomic number.

pH A measure of hydrogen ion concentration (acidity). pH = $-\log[H^+(aq)]$

Polar molecule A covalent molecule which contains atoms with different electronegativities. The electron density of the bonding electrons lies towards the more electronegative atom.

Polymer A large molecule formed from many smaller monomer molecules reacting together.

Polymerisation The reaction of monomers to form polymers.

Precipitate An insoluble (solid) product formed when two solutions are mixed.

Proton A positively charged particle found in the atomic nucleus. It has a mass of approximately 1 atomic unit.

Radical A species which has an unpaired electron available for bonding. A radical is formed by homolytic bond breaking.

Rate of reaction The change over time of the concentration of a reactant or a product of a reaction. Its units are mol dm^{-3}s^{-1}.

Redox reaction A reaction in which oxidation and reduction both occur. One species is oxidised, while the other is reduced.

Reducing agent An element or compound that loses electrons to an oxidising agent, which itself gains electrons in the process.

Reduction A process in which a species gains electrons. It can also be defined as a decrease in oxidation state for redox reaction.

Relative atomic mass (symbol A_r) The mass of one atom of an element compared with one-twelfth of the mass of one atom of carbon-12.

Relative molecular mass (symbol M_r) The mass of one molecule of an element or compound compared with one-twelfth of the mass of one atom of carbon-12.

Reversible reaction A chemical reaction which can take place in both directions and so is incomplete. A mixture of reactants and products is obtained when the reaction reaches equilibrium.

Salt A compound formed when an acid reacts with a base.

Saturated organic compound An organic compound that contains only single bonds between the carbon atoms.

Shell A term that is sometimes used to describe the principal electron energy levels in an atom.

Structural isomers Isomers having the same molecular formula, but different structural formulae.

Substitution reaction A reaction in which an atom or group forming part of a molecule is replaced by a different atom or group.

Unsaturated organic compound A compound that contains one or more double bonds between the carbon atoms in its molecule.

Van der Waals' forces A form of intermolecular bonding. They are forces between temporary dipoles in adjacent molecules. Van der Waals' forces are between one-hundredth and one-tenth as strong as typical covalent bonds.

Index